Lothar Späth

Facing the Future

Germany Breaking New Ground

Springer-Verlag
Berlin Heidelberg New York
London Paris Tokyo

Lothar Späth
Minister President of Baden-Württemberg

Translated by

Veronica Leary
Franziskanerplatz 8
8400 Regensburg, FRG

The German edition was published under the title
"Wende in die Zukunft"

ISBN-13: 978-3-642-71610-2 e-ISBN-13: 978-3-642-71608-9
DOI: 10.1007/ 978-3-642-71608-9

Cover design: W. Eisenschink, 6805 Heddesheim

Data conversion, Offsetprinting and Binding: Appl, Wemding

2142/3140-5 4 3 2 1 0

Foreword to the English Language Edition

When Bruce Nußbaum, co-editor of "Business Week," prophesied Europe's inexorable decline in his 1983 bestseller "The World after the Oil," there were not many people around to stand up and contradict him. The political short-comings, the technological backwardness, and the defeatist attitude of the Old Continent, which only seemed to be preoccupied with the past and itself, were all too obvious. In a number of speeches which I gave on the East and West coasts of the United States of America at the time, I repeatedly raised the question: "Has Europe run out of Steam?" For the majority of my audience, however, the matter had already been settled, by Nußbaum. When I attempted to expound on why I did not take this pessimistic view of the matter, I was met with reactions ranging from polite astonishment to frank scepticism. European shares were not exactly at a premium at that time.

A number of things have changed since then. European companies have been able to catch up with the world leaders in high technology research, development, and production much quicker than many people would have thought possible. A strong and steady economic upswing, together with a – for Europe – advantageous rate of exchange for the dollar, has led to enormous successes in exports and high surpluses in the balance of trade and production for the Federal Republic of Germany and a few other European countries.

This upswing has been noted with interest in the USA, but the opinion is still very widespread that it is entirely due to the exchange rate which has been having a negative effect on the American economy for some time now. At best, among experts one speaks with respect about the efficiency and ability to adapt on the part of European management. These are doubtlessly important factors. By themselves, however, they do not provide an adequate explanation for what one could euphemistically refer to as the

"European renaissance." The most important factor contributing towards Europe's newly awakened self-confidence are the Europeans themselves.

Europeans have begun to actively face the future. At the same time, they are not forgetting the values and achievements which have made the continent what it is today. They believe that there is no reason why European research and technology should not be as good as what the Japanese and Americans have. This attitude is especially widespread among the younger generation, and a youth that believes in its future is the most valuable asset a nation can possess.

"Facing the Future" was written when the force of the new changes was also just beginning to be felt in the Federal Republic of Germany, without it being apparent whether they would end up succeeding or not. The main purpose of my book was, and still is, to encourage and support these forces. This called for two things: First of all, past mistakes had to be openly admitted and analysed in order to learn from them. Secondly, it was necessary to describe as concretely as possible the qualitative standards involved in the transformation from an industrial to an information society and to express them in the form of perspectives. For the European industrial nations, namely, economic growth and technological progress alone will no longer be enough to satisfy the needs of citizens in the future for public and personal models.

This led to a vision of society which is capable of reconciling the demands of technological progress with its social and cultural roots. This model of society cannot, of course, be generalized because it is based on the concrete historical situation and actual conditions of Germany. "Facing the Future" was written primarily with the German reader in mind, who is familiar with conditions in his/her native country.

The idea behind translating and publishing an English language edition necessarily involves other criteria and intentions as were originally conceived in the writing of this book. In particular, for the American reader this edition might convey a sense of the change in attitude which has taken place in the Federal Republic of Germany within the period of just a few short years. Anyone who wants to know why Nußbaum's end of Europe scenario has not yet taken place, and in all probability never will, might be able to find some answers in this book. As an engaged and convinced "transatlantist," I would be particularly pleased if this book could also make its own small contribution towards making the sometimes complicated German way of thinking, which is often difficult for outsiders to understand, more transparent.

August, 1986

Lothar Späth

Foreword to the German Language Edition

One cannot help suspecting that politicians who write books either have time on their hands or a false conception about what their job is all about. Is not politics, more than any other field of activity, endowed with the power of being able to realize one's ideas and initiate changes? Why bother taking a detour with the written word? In view of the profusion of published material, no one would seriously miss the politician's literary contributions. The politician should act, making arrangements and carrying through concrete measures – there are enough other people who have made it their business to reflect and provide commentary on the state of affairs.

Nevertheless, the more experience I have with this allocation of roles, the more dubious it appears to me. The "power" of democratically authorized politics in a free society is, in truth, quite limited. It consists more often in a pretence of power, corroborated by political enemies in order to discredit one, or by political friends to enhance one's success, than in a substantial and decisive influence. Initiatives to enact new laws quickly come up against the resistance of other areas of responsibility which are equal or superior to one's own sphere of influence, at the state, federal, and European Community level. Many administrative decisions do not create the necessary clarity for either those who make them or those who are affected by them, due to the uncertainty of the judicial screening processes, which are generally involved and drawn out. Overnight, sensational new items, ranging from the discovery of toxic wastes at dumping sites to investigations into the private finances of people in public life, appear in the headlines of the press to change and decide the day-to-day course of political activity. Political decisions in many areas seem to be simply derivations of preliminary, scientific decisions, for example, as far as environmental protection, research support, new highway routes, airport expansions, or nuclear power plant sites go. In most cases, the political scope of action tends to amount to zero.

Our society has established a number of controls for limiting the development of political power. The separation of powers is the most traditional factor, the freedom of the press the most effective, and the increased expectation on the part of many citizens to play a more active political role will probably prove to be the most decisive factor in the long-term, to the extent that it makes optimal use of both the guarantee of legal action and the possibilities for shaping public opinion.

From the point of view of a democratic form of government, there is nothing to object to in this. Things only become difficult, even self-contradictory, if one expects politics, with its complicated and time consuming feedback processes, to function like some kind of ideal mechanism where you simply have to push the famous button and "presto," you get instant results. It is often, however, the very same citizens who criticize indecisive politics who appeal against sovereign decisions to the highest courts, if they themselves are the ones who are affected. It is the very same press which calls for more executive decision-making power in the environmental area and more social justice in individual cases which complains at the same time about the bloated government apparatus.

My purpose is not to point an accusing finger, to try and shrug off the responsibility for the lack of political activity onto someone else. A society has to have some idea at least, however, of what it wants and which values to give priority to. If you subject each step in the process to controls, you cannot expect the same rate of production as you would achieve if you were simply satisfied with an interim and final check. A river with many rapids is certainly ecologically more valuable than one that has been artificially straightened out, it just is not navigable. By making one certain aim absolute, other aims and purposes are inevitably excluded.

My impression is that this fact, as simple as it may sound, has not really been registered by the public and embedded in their consciousness in the Federal Republic of Germany.

On the contrary, there is a "split" in the socio-political discussion along the following lines: Associations, interest groups, and leading journalists are responsible for formulating the aims and demands. Taken together, they cover just about everything that the public might expect to want in a modern industrial country – a healthy environment, humane technologies, new and secure jobs, social justice, less government, a reduction in the public deficit, more public and private investments, better infrastructures, more direct contact to citizens, a more efficient administration, the smallest possible amount of data capture.

The responsibility for steering towards and achieving these aims – without really giving much thought to the matter – is automatically assigned to politics since this is, after all, what politics is for in the first place. Politics also has the necessary "power for pushing things through" at its disposal.

Politics, in its turn, is more than willing to take up the myth – it is the only way of mobilizing voters to the polls. The tendency is to develop a whole roster of control mechanisms and to leave the articulation of more global social aims, for the most part and without resistance, to opinion makers outside the sphere of the direct political decision-making process. First of all, this is easier, secondly, there is no time for "philosophizing," and thirdly, concrete activities are the only things that count as proof in public life.

Predictably enough, the result of this "splitting of areas of responsibility" is fatal. The gap between the largest common denominator of spelling out demands without, however, taking on the political responsibility for their realization, and the smallest common denominator of what ends up being carried out is continually growing. The accumulated frustrations lead to further polarization. On the one hand, we are continually being confronted with urgent reminders of the aims we should strive for and the negative consequences of not achieving these goals; on the other hand, there is a tendency to try and pass off every small step in the right direction and marginal improvements as the big breakthrough towards "fulfilment." At the end of this spiral there is the threat of the loss of credibility, an increase in uncompromising fundamental criticism, and the fragmentation into political cultures which are no longer capable of communicating with each other.

I am not claiming that we have already arrived at this state of affairs. No one can deny, though, that there are disquieting tendencies which point in this direction. In this kind of situation, the politician should not leave the public debate concerning global aims entirely to organized interest groups or the media. The politician has to try and develop his/her own perspective, even at the risk of it being a step ahead or even miles beyond his/her current radius of action. He/she has to promote understanding for the standpoint that in a pluralistic society the sum of its parts always adds up to less than the total. If we try and compress every aspect of a certain problem into a self-contained system, without any freedom for comparative, qualifying, or, in terms of a compromise, reconciliating positions, we will end up with an ideology.

My main purpose in writing this book is to take up a position which is comprehensive and open to the opportunity of establishing a "reconciliation" between counterpoles. I am motivated, above all, by what I have already pointed out as the disquieting effects of the growing one-sidedness and, at the same time, growing claims of absolutism, of the political arguments in our public debate. We make our aims absolute, stick stubbornly to our own standpoint, reject solution models as imperfect here in the Federal Republic of Germany, while the world all around us is growing economically and technologically together at a breathtaking pace. Nations are formulating long-term, joint perspectives, incorporating a broad consensus of government, science, the economy, and society in order to achieve these aims. Companies are working on a world-wide basis to establish communication and information networks which, in the not too distant future, will be more important in terms of infrastructure than national highways and transportation means. New axes of cooperation, above all in the Pacific region, can be seen which conform to the law of economic power and which could result in shifts of political power on a global scale.

The shorter the interval is between economic and technological changes, the more need there is for strategic planning of aims and developments. Strategy, however, calls for a high degree of organization and the willingness to compromise and cooperate. This cannot be achieved in a democratic society by orders from above, but has to be done by patiently trying to convince people of the rightness of certain policies. Whether we will manage to accomplish this is not a question of political or social methods; it depends mainly on the credibility of our models, together with our basic convictions.

In the past the Federal Republic of Germany has got along without the motivating power of social models. Having a lot to catch up on in the way of material goods on the part of individuals and the free enterprise system as such were adequate stimulants. The lack of perspectives in our society, however, has become obvious and detrimental to our future development with the growing problems that we are being confronted with regarding our structural adaptation, the immense shift in values between the generations, and the inflexibility of a bureaucratic distribution system.

We have to start by facing our future together. This means parting company with our outmoded industrial society, not only economically, but also politically and intellectually and judiciously paving the way for the new territory that lies ahead of us. This takes time, it cannot be accomplished overnight. Our path is mapped out for us, however, and there is no turning back. Retreating into the past would mean throwing away our future. We have options, however, which we have to recognize and take advantage of. The industrial society will be followed by a new form of society which is usually (in this book as well) referred to as the information society. More important than terminology, however, are the goals that are seen in connection with the new society and the will to form them accordingly.

The change in technological structure, with its political, economic, social, and cultural repercussions, forms the core of this book. This comprehensive development has not reached its final phase by any means, so that a number of statements have been made realizing that they might soon have to be reexamined and revised in keeping with new developments. Between the time I began writing this book in the fall of 1984 and the delivery deadline for the manuscript, a number of changes had occured: Public debate had not been centered anywhere nearly as intensely on the the new technologies, as it is now – fortunately and justifiably – the case; realistic considerations of employment and adaptation were not in the forefront, but polemic and pithy Orwellian visions.

I was surprised at the one-sided approach that many important interest groups in our country took in regard to a social problem that will be a decisive factor in shaping our future, and I have attempted to look into the causes. Although things have doubtlessly taken a turn for the better in the meantime, and scientific and technological, as well as economic questions of structure are being treated with due earnestness, I have let stand what appeared to me at that time as worthy of criticism as "Notes on the Culture of Politics." In retrospect, the matter no longer appears as urgent to me as it did a year ago, but it is still not without validity.

The other chapters also bear witness to changes, to the investigation and assessment of matters. New impressions, new insights, and further questions were continually cropping up. The whole book is therefore more of a "transcript of a learning process," than a conclusive assessment and judgement of established facts. In the midst of the current of radical changes, matters could hardly be otherwise.

Lothar Späth
Stuttgart, West Germany
August 1985

Table of Contents

Notes on the Culture of Politics

Chapter 1
Exhausted Energy

The Rise and Fall in Public Concern

The fact that one topic can catch the public's fancy for certain periods of time, sometimes to the exclusion of everything else, is one of the basic principles that seems to be part of public life of a nation. Like the rising crest of a wave, public debate is concentrated on one topic of general concern to such an extent that within a short period of time the urgent need for political action inevitably arises.

There is no uniform pattern as to how such "national cycles of interest" come about or how long they last. The educational reform movement which caught on in every state in the Federal Republic of Germany during the mid-sixties was sparked in the main by a series of articles, "Die deutsche Bildungskatastrophe" (The German Education Catastrophe), which were written by the educational philosopher Georg Picht in 1964. Previous to this, isolated publications had dealt with alleged and actual shortcomings in educational policy. It was Picht's forewarning, however, which broke the ice for the implementation of concrete political measures. He pointed out that there soon would not even be enough secondary school graduates to enable the requisite number of teachers to be trained for the next generation, to say nothing of providing an industrial nation with a sufficient reserve of higher management personnel.

The planning and centralization zeal which spread like wildfire in the early seventies, on the other hand, received its crucial impulse from government officials. It was inspired by the social sciences which were at the zenith of their influence and declared that a model structuring of reality was an indispensable instrument for any kind of responsible political action. The movement was also animated by the amazing increase in the number of planning offices in the administration in Bonn, and by the state governments who all rushed to immediately rectify "negligences" in their hitherto existing plans for the future. Community reforms, district reforms and functional reforms followed one another in rapid succession and tied up considerable political and administrative resources over a number of years. To be precise, the administration for many years was, for the most part, preoccupied with itself. Since this corresponded to the general spirit of the time it did not seem to strike anyone as amiss and was considered unavoidable.

A further pronounced cycle, a topic which is still riding high on the current crest of public concern, is the ecology movement.

The theme of environmental protection can be found in the platforms of every political party and in the writings of many leading economic theorists as far back as the sixties. As early as 1962, for example, Alfred Müller-Armack pleaded for "an improvement in the public environmental factors of our life" as the focusing point of future economic and civil policy. He called for a second phase of social free enterprise established on ethical and philosophical principles.

Billions were spent, without sounding any special public echo, on systematic sewerage purification and comprehensive refuse disposal – an accomplishment which can only be fully appreciated by frequent travellers, who have to learn to do without such niceties of civilized life. It was not until the Club of Rome published their startling, if somewhat controversial, report about the limits to expansion in 1972 that the environmental movement got off the ground and a new cycle was launched.

There are a number of factors contributing to the current predominance of the environmental issue – objective reasons as well as ones stemming from politics, intellectual history and economic development. There is no need at this point to go into detail concerning these factors since our main point of interest lies with the oscillatory course of public "states of agitation." Later, in connection with a comprehensive analysis of the new social movements, there will be all the more to say with regard to these factors.

Conceivably the swing of the pendulum, the fate of every political movement, will some day also catch up with the ecology debate: after an awareness phase in which the pressure of public opinion mounts and demands for the rapid realization of political measures increase, the pendulum of interest slackens and a counterreaction sets in.

The educational reform of the sixties led within a short period of time to a number of radical changes. Standards for secondary school students were downgraded and university entrance requirements lowered. The universities had to adjust themselves to academic training on a wholesale basis, and there was a considerable increase in the number of faculty members and administrative personnel. Finally civil servant positions had to be created for the thousands of young teachers and administrative lawyers taken on by the government every year.

The last link in this chain of events had barely been formed when the counterreaction had already begun. The evident drawbacks to the reform concept drew heated criticism: a loss of quality in education, positions blocked for talented young scientists, resignation and a brain-drain in the field of research, an inflated bureaucracy, a rapid increase in the federal debt, and a decrease in occupational opportunities.

Twenty years later we are now almost back at the starting point as the urgent call for an elite corps of talented young to secure our future once again sounds. We are now trying to meet this demand by starting over again. This time around, however, the motto is "class instead of mass" – and we no longer have the financial and the structural scope of action which we possessed two decades ago.

A similar trend can be seen in the administrative reorganization which the "powers above" set in motion. The alleged increase in efficiency had barely made itself noticeable, to say nothing of saving anything in terms of dollars and cents, when adjustments and concessions had to be made in deference to change in public opinion: villages and subdivisions received a greater voice in making community decisions, new intermediate levels were created for voting about neighbourhood and community affairs, free school busing was the costly consolation prize for the closing of village schoolhouses, municipal centers were badly in need of extra financial allocations to cope with the burden of incorporating smaller towns into their district. Today no one can, or very wisely wants to prepare a balance sheet of costs and returns of the administrative reform. The cultural, social and psychological about-face of the population, on the other hand, cannot be overlooked and can be seen in every town and village and on every community bulletin board: decentralization, small is beautiful, the cultivation of limited, assessable spheres of action. The citizens themselves have in their own way taken care of counterbalancing the results of an all too trusting faith in public trends.

The cyclical preference given to certain themes or a certain perspective is one of the distinct dangers facing the future of economic growth and political development in our free society. What at first glance appears to be an absolute imperative, supported by those in the know and the press, usually turns out to be the ideology of a limited viewpoint and the circumscription of our sense of reality. What seems to be simply a matter of common sense, uniformly urged upon the public by the media, special interest groups and panels of experts, is in reality generally the first and decisive step towards polarization. Political measures are put into force whose excesses later have to be pared down at enormous costs in terms of money and friction.

Above all, though, what happens to those developments which are not "lucky" enough to get caught up in the stream of general public interest to ride the crest of popularity? How are subject matters dealt with which do not happen to enjoy the goodwill of the media, topics which are considered inopportune and which lead to vocal criticism of the politicians who take up these issues?

Such developments simply do not materialize – at any rate not until they manage to hold their own ground against the opportunistic pressure bearing down on them by the problems of current immediate interest.

During the seventies, for example, a renaissance of research and technology, even remotely comparable to the breathtaking strides being made in the USA, Japan, and the countries of Southeast Asia did not take place in West Germany. The Federal Republic was blind and deaf to the signs of fundamental change underway in the scientific, economic, and technical fields in the world around it. There was no political or medial force to counterbalance the homogeneous interpretations and explanations of the social sciences, which dominated the public view of things and imputed a danger to society in any scientific or technical change.

This led, in hindsight, to a monstrous misjudgment of the international economic development and the appropriate course of action to follow. Between 1976 and 1982 federal funds for data processing, micro-electronics and automation technology were cut by more than half. At most German universities the budgets for personnel and equipment in the computer science field were drastically reduced. The architects of higher education felt justified not only by the general political culture which granted the social sciences priority; they were justified to an even greater extent by the "latest" labor market policy forecasts of the Federal Department of Education which predicted a large surplus of electronic engineers while at the same time the demand for metallurgy and mining engineers was to remain constant.

In 1972 this same department did, however, publish the translation of a study by the Committee for Technological Progress of the Japanese Board of Economics, which advises the Japanese Prime Minister in matters of economic and technological policy. The study was a joint venture by Japanese experts in the fields of science and industry and gave top priority to the following areas of research especially deserving of government funding: space exploration, oceanography, atomic energy, the development of computer technology, increased support of industrial research, and an intensification of biomedical and environmental technology. The study included a list of recommendations with the title "The Seven Keys to the Future" in which the authors call for the creation of think tanks, new forms of international cooperation, the establishment of efficiency control systems and the training of a large reserve of scientific and technical personnel.

As mentioned above, this was in 1972, the year in which microprocessors first began to be sold in large numbers in the USA and the term "Information Society" was placing its stamp on the Japanese national view of the future. In the Federal Republic of Germany, on the other hand, east-west politics and social topics dominated the public scene.

The point is not to rehash party politics, but to elucidate the decisive factors which led to the fundamental international changes indeed being put on file in the different government departments but without anyone taking particular notice of them. There was no political, no journalistic and consequently no public interest in questions pertaining to international trade and economics.

Quite the opposite: In view of the exaggerated scepticism towards industrial expansion, which was bolstered by the Club of Rome's report and the belief that the correlation between the dynamic force of economics and the job market could be neglected, any call for an increase in efficiency would have been seen as an anachronism, as the reactionary resistance on the part of the technocrats. University professors could not have lectured without harassment or politicians been able to show themselves in public, judging by the current barraging of the press.

The last decade has taught us that almost no one is in a position to escape the pressure to conform to a certain way of thinking, which is the result of a cyclical concurrence on the part of an opinion making elite. If this trend is directed "within," there is a preoccupation with the social conditions of the self and the "without" does not exist.

In the collective infatuation with our own mirror images we are simply marking time. Dynamic forces come to a standstill. The propeller of economic development, which in an exporting country is kept going by the external incentives of an adequate edge in competitive trade, starts to slacken. The blades continue to rotate for some time, due to the large mass of substance – accumulated capital and goods – and a certain degree of self-dynamic at the disposal of the economy. But at some point the loss by friction outweighs the waning influx of energy from outside.

In the meantime, we have learned to recognize this point as a result of our unfortunate experience. For an industrial country the size of Germany it will take about ten years for most of the substance to be exhausted. How long will it take before it can be renewed?

No one can say for sure. It is also not the issue at point. On the contrary, the important question is how we can change our country, West Germany, so that the future does not pass us by. It is a question likewise posed for the politicians, the political parties, the different associations and unions, and the media.

Chapter 2
Politics as the Fiduciary of Future Generations

The Role of the Political Parties

The susceptibility to cycles of excessiveness leads, as Kurt Sontheimer aptly pointed out, "to a trend to pursue extremely short – term policies that – though with an eye to the future – still do not manage to produce anything more than patchwork policy in keeping with the dominant spirit of the age currently in vogue."

Ambitious policy making of a higher caliber must attempt to avoid causing the pendulum to swing in the one or the other direction. The onesidedness of many standpoints and procedural patterns basically demonstrate a fundamental failure to establish a conclusive overall picture in which individual aspects, according to their status and sequence, can be integrated. It is, therefore, not just uncritical, political hangers-on, those voting opportunistically according to the way they think the political wind is blowing, who are harmful. Even a strictly anti-cyclical policy, if it had nothing else to offer, would not have any more influence than the next counterreaction within sight.

This is precisely the point that is often so difficult to get across. If a certain measure has proven to be misconceived, nothing seems more natural than to have the matter immediately corrected. If an opposing political party, whose hitherto existing role involved maligning current government policy as much as possible and convincing voters of the wisdom of its criticism, comes to power it is under tremendous pressure to do everything differently. The "how" of it, because often complex, is of secondary interest. Quick results are supposed to serve as proof of a successful change of course. Anyone who counts himself as being a victim of the old politics hopes from the new officeholders, whom he helped to power with his vote, to have his individual expectations fulfilled as quickly as possible.

Politics which attempts to provide a cure-all for the social evils afflicting different groups on a case to case basis is doomed to failure from the start. It ends up doing an about-face with a throwback to the cyclical excessiveness of its predecessors due to the transilience of its measures. New disadvantages are created which provoke the pendulum of public opinion to swing back in the other direction. Involuntarily people are drawn to those who are in a permanent position of opposition and whose alternative is the promise of but never the attempt at fulfillment.

This shows that continuity in policy can only be achieved on a long term basis if one reasons along solid public policy lines and not by doctoring around on isolated symptoms, if sound social politics and not group interests serve as the models for determining spheres of action, if categories of objectives in their entirety and not isolated political domains form the conceptual foundation, if, in short, providing for the future carries more weight than satisfying present interests or overcoming the past.

Does this mean that we are asking to square off the circle? It might seem so. One should take care, though, of regarding one's own shortcomings as the inevitable impediments of nature and unforeseen miscalculations as ineluctable law. Resignation is sometimes nothing more than an overestimation of one's own shortcomings, an infatuation with the underdog position. After sunning ourselves for decades in the bright light of the German economic miracle and feeling like industrial muscle men par excellance, we are now suffering from a hangover which has taken on the form of a national disaster – another unfortunate exaggeration which needs to be put into proper perspective.

One associates public policy primarily with a certain standpoint towards economic and social political issues. However, it is a question of conclusions based on foregone value judgments which go much deeper and which directly touch upon the conflict between the human condition and the constitutional organization of society, upon the individual and general ethics. The demarcation between self-fulfillment and integration is a standing issue of controversy in the modern democratic state and its socially satisfactory synthesis a still unsolved problem for the future.

This antithesis is less astonishing than at first appears if one recalls the youthful history of modern democracies, which spans just over a hundred years and whose cornerstone can be considered the establishment of universal suffrage in free and secret elections. The concept of the democratic constitutional state is primarily the concept of the limited power of authority by means of a division of power among different responsible bodies and by a non-arbitrary organization of political decision making from the "bottom" to the "top." Civil rights were conceived of as a breaking away from regimentation by the state and acts of arbitrariness and violence and not as a move towards more community orientated obligations. It was not democratic state theorists but rather the founders of the Christian social doctrine which attempted to bridge the gap between individual merit and the idea of community responsibility. A system of decentralized, subsidiary social responsibilities ("Subsidiaritätsprinzip") and solidarity are social classification concepts which have their roots in the Christian perception of the human being and not in any constitutional rights or duties called for on the part of democracy.

The established political parties so far have not been able to form concrete spheres of action for comprehensive public service which go beyond an understanding of their traditional role as anything more than a rung in the ladder of the political decision making process. The constitutional status assigned to the parties as mediators of the public will has been more of a liability than an asset for development along these lines. Parties as upholders of the state are preliminary forms of the state itself, their officials are political professionals; public motivation occurs primarily in periodically mobilizing supporters to the voting polls.

A good deal of the fascination of alternative movements is due to their contrast with established parties. They will attract followers as long as they can manage to satisfy the need for involvement and creative impulses with imaginative and symbolic grass-roots activities. Not the result, but the shared experience encountered is crucial.

Are the large democratic parties really incapable, in their turn – and without imitating a still rather infantile actionism – of initiating a perspicacious discussion which could involve the individual, stimulate commitment and provide participatory enthusiasm? I refuse to accept a defeatist standpoint which considers such ameliorations to be beyond our reach for reasons inherent to the system. On the contrary, I believe that it has never seriously been attempted. Day to day politics is preoccupied with routine administrative work, regulation politics is engrossed by superficial disputes, and discussions about image and fundamental values take place beyond the closed doors of commissions and the drafting committees for party platforms. The stress on perfectionism, the need to present the voting public with a ready-made view of life and to limit controversial discussion to particular measures and steps while avoiding (the naturally much more vulnerable) political visions, opens the door to the desolation of political culture. The future is divested of political consciousness, secondary topics of current interest are aggrandized to questions of vital national concern, matters are superfluously polarized while essentialities, things it would really be worth arguing about, are not gotten around to.

What are the consequences for someone working in the practice? One example: politicians should not see themselves above all as administrative executives – they have well paid civil servants for this function. They should consider themselves to be just as much the trustees of future generations as they are the agents of the present one even though, or rather for the very reason that the former do not yet have a voice. They should spend less attention to petty matters which citizens in their own sphere of experience are adequately able to deal with without government interference and more heed to intercorrelations, changes, networks and trends. Since they themselves are not in a position to accurately assess such numerous factors, they need professional guidance. Not biased, specialized groups are called for, which compete with the official bureaucracy, but rather interdisciplinary teams with an active interchange of ideas and mutual feedback. It must be possible, namely, in the age of computer simulation to prepare material dealing with demographic, sociographic and economic trends and cycles so that situations of scarcity or surplus are indicated in good time. In fact, however, it is tremendously difficult to obtain interdepartmental or intersectional data which could reliably indicate the effect of a single development on other branches and what a long term decision for other political fields would mean. In Baden-Württemberg we are presently involved in developing a computer supported information and communication system for the entire state administration. Our greatest difficulty, that is already apparent, will be to find enough qualified personnel who can work "management orientated," that means being able to deal with administrative and operational matters on an interdisciplinary basis. There are data processing facilities that have been in operation for some time now in a number of administrative departments, but the exchange of stored information and particularly its coordination on higher administrative levels leaves much to be desired. As a result it is extremely difficult, for example, to obtain detailed information about the current budget enforcement quickly and reliably. It will probably take a number of years till we will have established a fully automated budget management on the scale that is absolutely necessary for a highly industrialized state such as Baden-Württemberg. We estimate, optimistically, that it will take ten years for the administration to undergo a complete conversion to a computerized information system.

Politicians should think first of all in terms of spaces of time and infrastructures and only after that plan in terms of deadlines and concrete projects. They should not be reluctant to design visions of the future or to take up ethical standpoints. Indeed, if they do not, they are robbing themselves of the dialogue with citizens on the most important political field of action. They have to, it is an inevitable part of being a politician, take the risk of personal failure.

Only such consistent politics, one with an open eye to the future, is in a position to provide continuity, to rouse public spirit, and to help determine the spirit of the age instead of drifting along in its wake.

Chapter 3

Fear of the Future: Cultural Survival

Crisis Among the Large Organizations

Germans seem to have a quenchless thirst for number mystics. When the first of January 1984 rolled around you could almost hear the sighs of relief escaping from the different editorial offices around the country: At last the portentous Orwell year had arrived; at last all the accumulated anxieties with regard to computers, the invasion of the private sphere, and surveillance could be pinpointed – 1984 was at hand. Scenarios conjuring up the ghost of Big Brother had been prepared well in advance and book publishers, feature writers, and TV editors alike simply had to pull them out of their desk drawers and add the finishing touches for the cameras to start rolling. The pessimists were in their element.

Those who saw a promise in the future also had their badge of identification, the number 2000. Book titles featuring this magic date usually focused their attention on aspects of micro-electronics, telecommunications and the "information society." Exposés packed with facts and detailed information provided a picture of the technical progress to be made by the turn of the century. The way in which factories of the future would function, and multifunctional LSI devices would change everyday life were among the main themes.

German scientists and economists started churning out one book after another in an attempt to bridge the gap between themselves and other countries which – politically inspired – had already made headway in getting a grasp on future developments. In Japan this was done by means of such studies as "The Information Society of the Year 2000" and "Japan in the Year 2000," in France by Simon Nora's and Alạin Minc's report "The Computerization of Society," commissioned by the French president in 1977, and in England by promoting discussion and declaring 1982 "The Year of Information Technology."

Our discusssion concerning the future was delayed in getting off the ground. This neglect, this long overdue neglect, though, cannot entirely be laid at the doorstep of the politicians. Other large and important groups within German society were also lax over the years in contributing to a constructive and worthwhile discussion of these pressing matters.

Social conflict monopolized the thoughts of social and political scientists during the sixties and seventies and there was no end to what they had to say to the theme. Educators and journalists also participated with enthusiasm in exposing even the most subtle forms of repression – the obscurer the better. They were less voluble, not to say taciturn, however, when it came to discussing the impact of the structural change in technology and its repercussions for society. A promising beginning was made in my opinion in 1983 with the committee report, "The Future Prospects of Social Development," which was commissioned by the state government of Baden-Württemberg. The report, which had the courage to also point out positive aspects of future development, immediately drew a volley of public criticism from certain quarters. The main theme of the Frankfurt Bookfair in 1984, "Orwell 2000," was also typical of the widespread tendency to fear the worst from new technology rather than to make the best of it. Even the last of the diehard optimists could no longer ignore the implication that the fear of Big Brother, if deferred, was to continue to be part and parcel of the intellectuals' baggage.

In view of a standing unemployment rate of more than two million and the undeniable threat posed by the "mechanical colleagues" it is no wonder that the unions are having trouble seeing things optimistically. Their negative attitude is also not apt to be changed by those recommending kill-or-cure remedies which would mean a negation of all the progress made by the unions since the end of World War II. The motto "pull up your sleeves, work harder, and be satisfied with less money" might appeal to a lot of self-employed people as the ideal solution, those used to the sixty hour week as a matter of course. It is not a realistic solution, however, because in a democracy in which various groups and organizations have an important role to play, their fundamental interests cannot be ignored without placing an undue stress on society.

From the point of view of sound economics there is also a great deal of doubt concerning the wisdom of such simplifications. An artificial reduction of labor costs would only alleviate the pressure to adapt to new technologies and thus create a decrease rather than an increase in productivity. An increase in the share of human labor in industrial production is not typical of any industrialized nation of the West, least of all Japan. The general pattern rather is to increase productivity by means of technology and capital investments thus making wage increases and the creation of new jobs possible. Of course, real wages cannot indefinitely be allowed to rise at a faster rate than productivity if increased unemployment is to be avoided. One should not overlook the fact, however, that in a modern enterprise the level of productivity nowadays depends more on the volume and use to capacity of physical assets rather than on the contribution made by human labor.

It is exactly this inexorable loss of the importance in employment generally which is the main problem facing unions today. The number of card carrying union members is stagnating, or in some cases clearly on the wane. One would be virtually at a loss to explain this development if one still held on to the idea of a homogeneous working class which perceived a protection of their interests exclusively in a united struggle against the forces of capitalism. For when, if not at times of crisis when people are exposed to relatively more in the way of personal risks, can one expect more sense of solidarity?

It is just as much an illusion, however, to talk about a homogeneous, working class consciousness among union members as it is to acclaim the alleged antithesis between capital and labor. The more personal and demanding one's job requirements are, the less apt an individual is to act according to the motto "power for the people." The clearer the connection is between the building up of capital, productive investment power, and job security of a company on the one hand, the more evident is the common identity of interests between employer and employee on the other hand. No doubt it is a fear of their own superfluousness in an individualized society which is currently causing the unions their biggest headache. New forms of behavior and spheres of interests are being developed which cannot be organized along traditional union lines.

This is no reason to rub one's hands. The outstanding strides made by the unions in this century have proved them to be a very important factor in counterbalancing excesses of our still developing industrial society. It is just over a hundred years since Bismarck's social welfare legislation in Germany prepared the ground for the first forms of health, accident, old-age, and invalid insurance. It was the unions and not the social democrats which formed the labor movement's focal point of interest during the time of the Kaiser and the Weimar Republic. Even after the socialist laws had been repealed in 1890, the unions had over 227,000 card-holders in 53 well organized central associations as well as 130,000 members in loosely organized alliances and local societies. Organized labor was thus not only more potent than the social democrats in terms of simple numbers but also in their power to call strikes and conduct labor negotiations. They were also willing right from the start, and in this lay their real strength, to risk losing face by compromising with those in power in order to improve the concrete situation of the common worker. Milestones in advancing the material prosperity of large segments of the population were the agreement to wage settlement contracts reached at the union congress in Frankfurt in 1899 and the refusal to participate in the general strike six years later in Cologne. An accurate estimation of the aims and means which the process of industrialization demanded of labor characterized both decisions. The union congress in Cologne in 1905 thus declared: "The congress holds a general strike as it is called for by anarchists and other people who are completely inexperienced in the field of economic struggle to be out of the question. Workers are warned not to be taken in by such ideas and not to neglect the small but necessary steps in their struggle for reform."

This was a clear-cut and forceful rejection of revolutionary and economically unfeasible utopias. Only seven years later, every fifth worker in Germany, more than two and a half million people all together, were organized in unions. The establishment of the unions as an integral and cohesive component of industrial society, as well as extensive social security benefits, have led to a twentyfold increase in the purchasing power of employees within this century. Since 1950 alone, real wages have increased fourfold in the Federal Republic of Germany – an unprecedented achievement which would be unthinkable without the fundamental moving force of "integration by means of limiting conflict."

The obvious changes being made in outdated industrial structures, however, will not be without repercussions for the unions and the way they see themselves. The interdependence between man and machine is mitigated by each new stage of automation. Actual working hours and operation times are becoming more and more independent of each other; more individualized work routines are being made possible and called for, not just in the service industries, but also in the area of production. In view of this situation can it really be in the interest of the unions to adhere to the fiction of standardized regulations for employee-employer relationships? Is it not about time to give greater voice to self-determination at work instead of clinging desperately to wholesale bargaining practices which, in fact, are becoming less practicable for more and more forms of enterprise?

The unions have to face the fact that it would be better to grant their factory committees more freedom and to satisfy themselves with the role of an umbrella organization, offering guidelines with regard to collective agreements concerning conditions of employment, instead of cherishing the myth of a collective industry and worker movement which no longer exists. The unions will have to pay more attention to those who want to master the new technology and show them how to achieve their aims. Collective bargaining agreements stipulating set periods of free time for further training, both on and off the job, would be a step in the right direction. Another promising approach for the unions would be to act as advisors and give support in negotiating contracts of employment tailored to individual needs and differing from company to company. Such contracts could enable the individual to choose the working hours best suited to the needs of his/her family and to influence working conditions according to inclination and interests. By recalling their own formula for success to memory, the unions should reach the conclusion that their position could be strengthened, rather than weakened by adapting themselves to new circumstances and demonstrating a change of consciousness.

Considering their technical capability and cultural importance the electronic media have also not made much of a contribution to a worthwhile discussion of the future. If we are really on the way to becoming an information society, a society, in other words, in which a majority of jobs are information orientated and the importance of information overrides that of raw material and concrete matter, we are faced with a strange situation. Television, the employer of information technology with the most widespread impact, has done very little so far in pointing out the path to be taken and goals to aim at.

This is not due alone to the nature of things or the medium, in my opinion, although one must admit that television has certain laws of its own which inhibit long term planning. Nonetheless, it was the electronic medium in Japan, for example, which took a lead in signaling the dawn of a new era and assessing the future as a challenge and opportunity. Economic relations, technical developments, and projects with long term objectives for society are popular discussion and information topics in Japan. Technological and economic planning, long term social problems, such as the threat of a disproportionate number of old among Japan's population, and analyses of Japan's standing in the international market are standard subjects of countless TV programs. With "television," a national vision of Japan as an information society of the next century is vividly illustrated. The population is being prepared for the new changes while the various fragments of information are fitted together to form a unified mosaic of the future.

Of course, one cannot just simply apply Japanese standards to the situation in Germany. To use the differences in culture and social structures as a ready-made excuse is not the answer either. No one is asking the public corporation networks in Germany to act as moral arbitrators in the classical sense of the word. Especially in view of the misuse of the media during the Nazi era, a healthy scepticism towards conformity in broadcasting is called for. One can no longer dispute the fact, however, that the new technology should also be part of German TV coverage. Even the English BBC, renowned for its high standard of journalistic integrity, has considered it necessary to broadcast introductory computer technology courses (in which more than five million Britons have taken part). Among German TV journalists, on the contrary, there seems to be a consensus of putting on blinkers when it comes to the important development of future technology. Prime TV time in the evening news is often set aside for demonstrations and protest rallies, while special reports are delegated to the less popular late night viewing slots.

While human chains, street blockades, and the storming of factory walls certainly make for a more dramatic TV news script, those kinds of demonstrations do not usually contribute towards a real solution to the problems. TV coverage of such "spontaneous" events exposes them to millions of viewers (for the sake of which they often take place at all) and bestows on them an importance which is usually all out of proportion to their actual significance. What is worse, however, and a matter of some concern, is the nurturing of the delusion that the malady of our times could be cured by such simple and, into the bargain, quite amusing patent remedies if only the powers above were willing.

But since the days of Max Weber nothing has changed to make politics less of a painstaking task. Particularly, as Germans, we should know better and be on our guard against gross oversimplifications.

On the other hand, one can hardly deny that the established political parties have been more than dilatory in creating vivid and imaginative models. Especially when dealing with matters of the future which are of general concern, plenary debates and government statements alone are not enough. There is an increasing need for direct, informative politics to coordinate and promote discussion, while showing the courage to adopt unconventional methods.

Only by cooperating can politics and the mass media provide an adequate answer to the challenges of the future, both in terms of what, and how matters should be conveyed. It is the job of the politicians to create more opportunities, public congresses, forums, exhibitions, and information events in which the worldwide race for technological leadership at least puts in an appearance.

New research institutes, foundation centers for young enterprisers and ambitious, science and technology projects are also ways of motivating and stimulating initiative for stirring up some of the restlessness taking a hold of other nations. For the last decade the Federal Republic of Germany has been a sleeping giant in the field of research and development and has already lost the advantage, which is all the more reason why politicians should be especially on the alert in this area.

The new political objectives necessarily have to go hand in hand with a new political understanding of communication in society. Otherwise, there is a danger of being misunderstood by the population in large. In the "National Development Plan," which the Japanese government drew up for a period of twenty years (1965 to 1985), the term "communicative society" continually crops up. It has become part of the everyday language in Japan, which in turn has made it easier for the media to identify with this new vision of society and to participate in its realization. The Japanese "Plan for an Information Society" even made provision for a comprehensive outline of legal guidelines to deal with the period of transition on the way to becoming an information society, although admittedly only a fraction of the plan in a few economic and technical areas was actually ever carried out. All the same, the goal of creating an open-minded climate for the change by describing new structures in political terms was fully achieved. Politics managed to awaken interest in the coming developments and thus was able to set a process of public discussion and action in motion which continues undiminished till this day.

There is no reason why a similar kind of procedure could not be initiated in Germany. A communicative society requires verbal intercourse, the search for knowledge and a questioning attitude. Politics should take a stronger stand about this than they have till now and not be afraid of undertaking both private and public "intellectual voyages of discovery" into the future.

By the same token, the media must also be willing to grant economic and technical matters their due recognition, especially with regard to their interrelationship with international developments. Particularly, if one recalls the way in which social conflict issues were handled, the eagerness with which such alleged and actual issues were once taken up, out of all proportion to the interest shown on the part of the majority of TV viewers, it is about time the program planners make up for things. Now that the majority of viewers and citizens are actually directly affected, their economic and social future is at stake, things should not be allowed to come to such a pass. Television, of all things, as an important information medium, should help towards contributing to an informed, rather than a disinformed society.

Let us return to our original starting point: We have said that a discussion of the future has been slow in getting off the ground in Germany because important factions within our society, for some time now, have been holding off. Whoever considers the examples named, the social sciences, unions, and television, to be one-sided along certain political party lines, should recall what was said about the way in which the established parties see themselves. The industry is also, at least as far as announcements made by their associations go, a long way away from offering anything that goes beyond considerations of business as usual. No real social alternative is being put forward which could motivate the individual to get involved in the "adventure" of the transition to an information society. The industry is in good company as far as their lamentations go, since just about every other kind of association and organization has joined in the chorus of complaints. The tendency on the part of industrial leaders themselves to indulge in pessimism, however, can only increase the doubts of their employees in the wisdom of adapting to the new technology and making the necessary effort to better qualify themselves. Paramountly, the future is a question of conviction concerning long term objectives, and not the accuracy of biannual predictions. Most business undertakings do not pay enough attention to this element of qualitative analysis. The result is a negative impact on dynamic change and on the willingness to take risks.

In trying to find a common cause of the conspicuous and widespread dread towards change, one cannot help suspecting that it is latent fears concerning one's own continued existence which have such a paralysing effect. And, in fact, nothing characterizes the new information and communication technology so much as its unbelievably rapid growth and the enormous increase in openings for new, individual options. Hundreds of thousands of people are already using video recorders and computer games to shape their own television programs. The German Federal Post Office is aiming to supply a million households with viewdata service by 1990, eventually evolving a much more autonomous type of consumer in the choice of information as well as products. The economy's need of obtaining essential information by direct access to a central data bank is growing, especially for noncorporate businesses and the self-employed. There was a sales boom in 1984 for home computers which could be had for less than a hundred dollars. The new electronic postal services, for the exchange of text material and graphics without using paper, although still in the initial stages of this technique, have already caught on in government offices and large corporations.

The trend is clear and irreversible. We are heading for the widespread use of individual computers with access to telecommunication networks which, after a phase of initiation, will be taken for granted . Many of the functions and services which are now carried out separately will be concentrated in a single device with access to a network system.

It is generally assumed that this process will involve more than just getting used to a few new methods of operation. What exactly the social and cultural consequences will be, however, remains unclear because of the traditional negative attitude of the social sciences in Germany towards technology. They are only just now beginning to discover the problem, and some organizations are still clinging to the status quo in an effort to protect their own internal interests.

The social scientists, as well as the politicians and established parties, though, will have to try and come up with some answers to the problem. Essentially, one is dealing in all three cases with the need to adapt to new circumstances. The trend is towards a greater freedom of choice and a more active role for the individual involving the urgent, growing need for guidance concerning the future.

The easy way out is to try and build "blockade cartels," in other words to take advantage of the widespread and understandable fear of changes and to put on new shows of solidarity. What this can lead to can be seen on hand from the British miners'strike in 1984, with its cycle of despair, confrontation and hopelessness. This is not the least of it, though. A free democratic society simply cannot tolerate having its hands tied at the cost of endangering future development, just because people would rather close their eyes to perhaps some heavy weather ahead and are having difficulty in getting a grip on and accepting new alternatives. To ignore the complex interrelationship of economic, social, and ecological factors under the delusion of some patent remedy, is not the answer either. There is no one cure-all for the world and never has been. Wanting to give the economy arbitrary and absolute priority would end up harming people, both physically and emotionally, just as much as the uncompromising ideologization of ecology would. By trying to play it safe and keeping ourselves high and dry, we run the risk of falling headlong into social disaster, whereas by taking measured steps forward we only take the chance of getting our feet a little wet.

Ideologists, those with a ready-made answer to everything, all have one thing in common: They have a poor opinion of the power of the individual and in fact, are afraid of it. At present, however, the threat of ideology does not seem as pressing as one might have feared considering the alarming situation confronting us in the late sixties and early seventies. The critical attitude towards the aims and values concerning the development of our civilization reflects the increased self-confidence and sense of liberation on the part of our citizens. On the other hand, we do not seem able to make the most of this trend and have failed in developing an integrated and progressive policy of social politics. Among the likely reasons for this are the lack of imagination, the exaggerated concern for and dependency on security, and the treadmill of everyday routine. The tremendous inertia of established institutions, however, is above all responsible. After rendering the Federal Republic of Germany great service in building up its industry, thanks especially to their cooperative strength, they are now finding it difficult to bring their own image into line with the new structures of self-organization, decentralization and flexibility.

Whether they like it or not, those involved in the political culture of our country, Germany, will have to face the fact of changes underway. In the degree that traditional topics of social conflict diminish in importance, the collective organs for checking such conflicts will carry less weight. Those interested in continuing their work in organizations such as the unions will have to come to terms with the increased need of justifying their traditional forms of representing group interests. They will have to assume that citizens will not just pay lip service to the new information opportunities, but that they also will take concrete and individual advantage of them.

We should welcome and not be afraid of this development. The information society of the future, however vague our idea of its concrete structure might still be, certainly will need one thing: informed citizens.

On the Way to an Information Society

Chapter 4
What Kinds of Shifts in Value will we have to Reckon with?

The Development of an Information Society

"We know very little about the social and political aspects of an information society." This statement by Karl W. Deutsch, the director of the International Institute of Comparative Social Research at Berlin's Science Center, is in remarkable contrast to a great deal of what one can usually read or hear about this subject. There is a growing tendency to already try and categorize a condition which we still have to come to terms with, to project a phantom picture of one's own fears and wishes, and to file away the future in neatly labeled pigeonholes. Information society conveys a number of meanings for different people. For the economist it is a new type of economic production, for the engineer a certain stage in the development of digital information and communication networks, and for the sociologist a qualitative change for people in the field of job occupation. While the sceptic sees it as a further step in the direction of a total technocracy, for the politician it is a "terra incognita." Depending on his/her viewpoint, one should warn about the writing on the wall or announce the inauguration of a new age.

Not surprisingly, there is a broad spectrum of publications offering outlines and hypotheses touching upon the possible social and political impact of the new technology. They range from "computopia," a kind of technical, social paradise on earth to "megamachine," the terrifying vision of a standardized society conforming to the information technology. Aurelio Peccei, the former president of the Club of Rome, considered micro-electronics "to be perhaps the greatest revolution in the history of mankind." Those who are best able to answer the information technology challenge, in his opinion, will shape and govern the new world civilization. This is exactly what computer critics, like the American Joseph Weizenbaum, are afraid of and is their rallying cry in their zealous crusade against the "Power of the Computer," the title of one of his books. The human being in all these accounts is depicted as drifting along in the current of technology rather than as rationally shaping his own destiny. One is faced with the alternative of either being a victim or making a pact with the devil. Damning technological progress as diabolical can only bring us to the precipice of a demonized, polarized, and, in any event, inhumane society.

Under the circumstances, a rational and moderate attitude seems called for, not as a means for procrastinating necessary political decisions, but rather, on the contrary, for preserving as much scope for action and control as possible. For the more we think we know, both in a positive and a negative sense, without this actually being the case, the greater risk we run of losing opportunities by rash commitments.

There is no empirical evidence dealing with the structure of an information society; there is sufficient material, however, concerning the process of our transformation into an industrial society, from its beginnings up to the present day, as well as adequate knowledge touching upon trends in the development of the new technology. If one pursues both the social and technological train of evolution into the not so distant future, the result is a framework in which the most likely and important characteristics of a future information society can be found.

Technical inventions, such as the mechanical loom and the steam engine, were among the most important, although not the only reasons for the medieval feudal society giving way to the first phase of industrialization. Other important factors paved the way for this development, in what must seem to us today to be an extraordinarily long drawn out process. There were scientific achievements (for example Newton's mechanics), philosophical currents of thought (such as the empirical philosophy of Hobbes, Locke and Hume, which can be regarded as the dawn of Enlightenment), social and cultural changes (the emergence of a bourgeois culture in towns and cities), as well as changes in business life (the accumulation of capital, the early beginnings of a division of labor, the theoretical foundation of Adam Smith). It was not technology by itself, but the development of technology under favourable social conditions which provided the impetus for overcoming the agrarian system.

The impetus, though, was powerful: From a majority of ninety per cent agricultural workers, who nonetheless could not prevent the periodic reoccurrence of famine, we now have a minority of only five per cent farmers producing considerable surplus thanks, among other reasons, to a tenfold increase in the fertility of the soil since 1800.

The gradual nature of this change could not prevent, as we know, giving rise at the same time to devastating new forms of destitution. Industrialization alone, however, was not responsible for the widespread impoverishment, as is often mistakenly claimed; poverty had always been there, it had simply taken on new and more visible forms. Mass emigration began in Germany after 1820, before the industrial expansion had actually gotten underway. The main cause was a rapid increase in the population (during the 19th century the German population almost tripled) along with the inability of agriculture to take on additional workers. The industrial society of early capitalism turned underpriviledged agricultural workers into just as underpriviledged industrial workers, till the steady increase in the number of new industrial jobs curtailed competition.

This process of restructuring took about a half a century, which was without a doubt a long and difficult time for the persons concerned. On the other hand, the foundations for decades of prosperity were laid at that time, which the descendents of these same early industrial workers were able to profit by and, up to the present day, still do.

It is worthwhile devoting some attention to the basic pattern as to how the reorganization of the economy and society by technology took place.

In the first as well as the second half of the 19th century a whole series of basic inventions found their way from tinker workshops into the factories. These inventions led to the establishment of whole new branches of industry, first in the area of textiles and heavy industry, later in the area of the electrical and chemical industries. A part of the innovations resulted in new products and caused a radical change in modes of transportation – railroads, steamships, and after the turn of the century, automobiles and airplanes. Without these new means of transportation, the industrial process would not have been able to achieve its widespread impact in creating new jobs and in establishing a network of international communication.

Another important factor was, moreover, the liberalization of economic policy, which opened the door for the development of enterprise by replacing old professional and mercantile trading regulations with the principle of freedom of trade. A change of orientation in state educational and information policy was also a very important determinant. Technical colleges, scientific academies and commercial advice centers (such as the Central Control Office for Trade and Commerce, which, founded in 1848, and under the leadership of its President Ferdinand Steinbeis, played an important role in the industrialization of Württemberg) were established and were essential in coordinating the changes in information and education.

Finally, the edge had to be taken off the new social conflicts, which occured as a result of the structural discontinuity in the economy. The innovative social policy of Bismarck and his social welfare legislation were able to channel and maintain a proper balance between the power elements of society.

The use of machines to replace and augment *physical* man power is generally considered to be the most important common denominator of industrialization. One should be careful, however, not to overlook two very important components of this otherwise quite accurate characterization: The supplantation of the raw material human muscle power placed a demand on natural resources – mineral resources and new forms of energy – to a degree that was unknown until then, and multiplied the number of negative influences on the environment. Man was only able to increase his physical powers at the cost of the organic balance of nature. Secondly, the endeavor to make optimal use of the different machines involved, in the end, not just radical changes in technology, but also had widereaching consequences for society as a whole. In this respect, the Taylor system of organizing work involved much more than maximum efficiency in terms of business management. Factory work now specifically became the link, an elastic buffer, between the different elements of a, for the most part, rigid machine system.

In this way, the value of human work in the industrial age of the twentieth century was increasingly determined by the degree in which it could be adapted to the larger technological system. One does not have to be a Marxist to recognize in this a fundamental process of alienation and isolation, less in terms of the work product as in the total spectrum of human abilities and potentialities, which were only allowed a very limited scope of action by the machines which determined the work rhythm. The human being increased his physical powers, not just to the detriment of his bodily well-being, but also at the expense of his natural talent for creativity and development.

In outlining a few of the important structural elements in the first wave of industrialization and the social and political changes it brought about, a number of urgent questions inevitably arise: To what extent can one compare this development to the structural change that is now becoming apparent? What are the most important criteria determining this recent radical change in technology? What are the advantages and disadvantages we can count on? How will the (rest of the) work performed by humans be organized in the future, will technology play a more important or less important role in determining the work rhythm? To what extent will the new technology call forth infrastructural, political and social innovations? Will we again have to overcome a period of hard times in the form of unemployment and disqualification, along with social rootlessness, before a new economic state of balance at a higher level of information and technology can be achieved? Last of all: With what shifts in values will we be able to, have to, reckon with? Is there reason to hope that we are getting closer to an ideal, free and humane society or are we moving in the direction of the kind of nightmare envisioned by Orwell and Huxley?

In the following, we will attempt to answer some of these questions. It is naturally impossible to predict the future with certainty, but we can offer well-founded hypotheses. Right at the outset, I would like to emphasize that in my opinion the need for social and political action will be much greater than many people seem to think, those who see the new technology simply as a technical development whose economic impact will be regulated automatically by the market. If that were the case, our calculations as to the structural dimensions of micro-electronics and opto-electronics, as well as to computer and robot technology, must be completely off. It would mean that we are not dealing here with fundamental changes in the economy, involving social and cultural repercussions, but simply with new products, like the thousands of others that appear yearly on the market. Everything points in the other direction, however, denying that such an oversimplification is an accurate assessment of the situation. Keeping the history of industrialization in mind, there is no way, therefore, of closing our eyes to the fact that our state and society have not simply been assigned a piecemeal, routine task but are being faced with a conceptional and comprehensive challenge.

A new infrastructure is again called for, zeroing in less on the problem of transporting people or freight, but concentrating more on the speedy exchange of knowledge and information. We need an economic policy which allows more leeway (especially concerning taxation), while providing guidance (particularly for the exchange of technology between public research centers and the business community). We need to establish an educational system which can provide a strong but flexible framework for building bridges between the public school system, job training, and further education. The clear objective must be set: enabling people to process information by promoting their talents, which go beyond the limits of information technology. A great deal of socio-political imagination will be called for to put a stop to the threatening inflationary growth of taxes and social contributions, set by law and the increase in the percentage of old people in the population. New forms of individualized social security must be found which are in a position to offset the decreasing contribution of human labor towards the G.N.P.

Only if we admit and develop a positive attitude towards the complexity of the problems, will we have a chance of coping with the transition to the age of information technology without considerable political and social dislocation, and while preserving our fundamental system of values. Those who want to try and take the easy way out, when things just are not as simple as that, only encourage ideologies. The appeal of ideologies has by no means completely died out, as Karl Dietrich Bracher has convincingly shown in his book "Zeit der Ideologen" (The Age of Ideologists). The latent desire for "simple truths" can easily give rise to powerful, new political currents, if democracy should fail to meet the challenge of the complicated and interrelated problems it is being faced with.

Chapter 5
An Intelligence Aid, But Not Intelligent

The Computer

The first phase of industrialization enlarged man's physical scope of action; by means of mechanical and electrical power he was able to multiply his muscle strength. The outstanding success had its price: The standardization of mass production, the foundation for the development of public social welfare, robbed work of a sense of its own qualitative value. In the span of two lifetimes more natural resources were used up than in the twenty preceeding generations.

The capability of machines to support and increase man's *mental* powers is generally regarded as the most important characteristic of the recent structural change. One often hears about an information technology of "intelligence aid" in this connection.

First, however, one must point out that, aside from information technology in the narrow sense of the word, there is also a broad spectrum of other technological innovations which will also have a considerable impact on the economy during the next two decades.

The problem of developing environmental protection technology, which is also economically lucrative, is one example. Its significance is quite comparable to that of the micro-electronic challenge and for a number of reasons. Aside from considerable research work, alterations in the economy on a large-scale and clear public policy objectives set by the state are called for. One of the major problems involved in environmental technology is the recycling of waste and other material, as is increasingly becoming evident. Entailed is also the question of an economic use of renewable forms of energy and how to store them, the development of which, till now, has been completely inadequate.

The substitution of metal with non-metallic material, such as plastic, fibrous and ceramic materials is of equal importance for the environmental and resource situation, as it is for keeping large, traditional branches of industry competitive.

The rapid, almost breathtaking strides now being made in the area of biotechnology will have an impact on a number of related fields before, perhaps sometime in the distant future, revolutionizing computer design. New plant strains and fertilizers will thoroughly alter agricultural production, modern drugs and vaccines will open up new fields of application in the area of medicine and pharmacology, all this at the same time giving rise to a number of difficult ethical and legal questions. Finally we have marine technology, the question of whose exploitation has already erupted in an intense struggle between industrial and developing countries, although still in its infancy, and the important area, in terms of infrastructure, of modern transportation systems. This brief survey alone shows that there is a whole range of technological developments which are just as significant for our economy and society as the information technology with its advances in the area of the media, office work and automated manufacturing.

The terms "intelligence aid" and "intelligent" are often used interchangeably. One hears about intelligent machines and artificial intelligence. One gets the impression that machine performance and the mental powers of the human are more or less interchangeable. Quite the opposite is the case. The value and importance of "brainwork" will increase, not decrease. Computers, in fact, can carry out so-called well-defined processes, which can be calculated one hundred per cent, with a precision and at a speed which are without competition. What we are dealing with here, though, is a limited, and by no means characteristic, component of the intellectual ability of Homo Sapiens. What distinguishes man above all is the talent to find his way in an unorganized world full of chaotic and unpredictable information by means of selection, association, intuition, calculated guesswork, experience and learning ability. With his sense-organs the human being registers about ten billion "bits," the smallest possible unit of data, in a single second. This information is instantly and simultaneously evaluated and prescreened by nerve cells so that only a manageable quantity of about 25 information elements per second are consciously perceived. The memory then retains about one information element per second.

What does this mean? The strength of the computer is man's weakness: his limited storage capacity. People have trouble in getting a grasp on and organizing their highly complex environment and are helpless when it comes to making autonomous decisions towards *shaping* this environment. Evolution has taught us that not being afraid of imperfection and risk is part of survival, and that it is generally more important to react quickly rather than spend too much time searching for the "perfect" solution. "The greatest asset of the human brain is its ability to offer a solution at all, whereas the computer at the most has options – programmed by humans – at its disposal," writes the physicist and computer expert Klaus Haefner, Bremen, in his book "Mensch und Computer im Jahre 2000" (Man and Computer in the Year 2000).

It is therefore very short-sighted, at least at the present state of affairs, to talk about "intelligent" computers or robots. To compare the most advanced computer models with the intelligence of a six-year-old, as one could recently read in a reputable computer journal, is absolutely irresponsible. One can only assume that the author has never run across a bright first-grader. Computers have been able to develop superior technological proficiency in areas that, in the course of evolution till now, have been of the least importance for the survival of the human species: the storage and retrieval of large amounts of data, the tireless repetition of simply structured processes whose information content is defined, reliable processing of defined programs based on mathematical-logical associations, but without any essential freedom of choice or power of decision.

Many jobs do not require much more than this in the way of qualifications, and are therefore especially endangered by rationalization measures. This is what makes the new technology such an extremely explosive topic, both politically and socially. It is one of the greatest challenges we are facing today. When measured against the course of evolution, however, and the standards by which humans see themselves, the subject ceases to be so dramatic and does not lend itself to horror visions. It is simply part of the continuous effort on the part of man to employ his unique intellectual abilities to delegate work to machines which is a burden to him, exceeds his own powers, or which involves personal danger. He is trying to create options for the future by directly compensating for his own shortcomings in the area of control and regulation. The human intellect is basically heuristic and the mind therefore does not function on a technical level. Human potential is being increased where it otherwise would be bound by the restrictions of nature. One could even go a step further and maintain that the very moment the complexity of the world begins to outstrip human comprehension and control, evolution takes care to provide an artificial aid. Since a biological adaptation within a short period of time is not possible, humans are provided with the intellectual tools to cope with the demands being placed on them and to avert the threat to their lives.

The idea of viewing the information society simply as part of the logical pattern in human development (being itself in turn, of course, just a transitional stage) stretching back to genesis, might provoke protest on the part of some. There is no use, however, in complicating transitional periods by mixing different aspects together and construing areas of fraction which ignore the valuable lessons history has taught us. Industrialization did not call man's right to existence in question, but rather forced him to reorientate himself in new fields of activity which brought advantages (a higher standard of living, social security, and more leisure) and disadvantages (a loss of work autonomy and family bonds). The information age is also not about to preempt man's existential rights, and will also involve advantages as well as disadvantages for mankind.

It is, of course, much more spectacular for the moment to adopt extreme positions. There is a widespread tendency to view things as either black or white.

Some people have interpreted the progress being made in the development of highly integrated computers, constructed along the principles of parallel processing of information, as a sign that man's exclusive realm of creativity is being invaded. They forget that the basic prerequisite for intelligent, and even limited creative activity, is the meaningful processing of complex impressions from the environment; even large-frame computers in the foreseeable future will not be up to processing all the information signals of a video camera if random objects in an alternating sequence of movements are filmed. Nowhere in the world have machines been built yet which can duplicate a cashier's "automatic" movements in handling and bagging merchandise. Some day such machines will certainly be built – which will then raise new employment problems. The question, though, is whether part of the human potential for *creativity* would thus be replaced. The cashier would be the last person to go along with such a view.

Chapter 6
Thinking in Black and White Terms is Not the Answer

The Expansion of the New Technology

Our work society tends to associate the social status of the individual with the kind of work he/she does. Work thus has become a way of classifying people, even to the extent of being the means by which we assess and judge our own worth. People think of themselves in terms of their jobs and their prestige as a reflection of society's recognition of their work. This narrow association between job and status comprises a dangerous standardization of one's purpose in life; especially in the case of unemployment the loss in job status can have disastrous consequences for the individual.

The humanities and social sciences should take on the responsibility for counteracting such oversimplified and restrictive value judgements. Unfortunately, they tend to do quite the opposite.

The position a person has in society is not determined by the optimum of his abilities as they might be able to develop under favourable conditions. A person's position, rather is made up of a "social cross-section," a role allocation which conformity to society, rather than individuality determines. It is this very stereotyping in determining one's social function, which is bringing us to the brink of a standardized world of automations in which people are victims, rather than active participants in creating their own living conditions.

The underlying reason for this phenomenon lies in the intrinsic nature of the social sciences and their fundamental shortcomings, which Friedrich H. Tenbruck impressively outlines. In their efforts to achieve scientific objectivity, the social sciences have succumbed to categorical labelling and generalizations. In the process, the individual, as such, has been pushed more and more into the background. A person's economic status, his social environment, all the alleged and concrete external influences at work, are seen as the decisive factors in determining an individual's personal state of mind. On the other hand, personal ethics, creativity, and the internalization of religious values have been ignored or at any event held to be so insignificant that the overwhelming force of one's environment is considered as easily outweighing their importance.

The idea of trying to take an individual's "measurement," using the sum of his social relationships as a yardstick, is widespread. The heteronomy of an increasingly industrialized work world corroborated the theory and, pars pro toto, it was generalized as the valid social pattern per se. Understandably enough, the unions and other large organizations had no trouble in going along with such a viewpoint. The emphasis on given "repressive" social patterns, which the individual can do little about, only indicated even more clearly the need for strong group solidarity to protect one's interests. More and more social science Phds are joining longstanding unionists from the rank and file in the upper echelon of leadership in many unions. There is even a trend towards replacing the realities of the working world with social theory, which can be explained by the close intellectual affinity between social science dogma, on the one hand, and wishful thinking on the part of the unions, on the other hand.

Neither is any help in coping with the social realities of the new technology. Man's intrinsic value is above all a matter of individual freedom of choice, and must be the basis for forming the rules of social integration. It is, in fact, in perceiving man simply as a derivative, a reflection of social classifications, that there is a real danger of relegating him to superfluousness. It feeds the fear of technology and brings the government into the arena as a shelter against life, while continually on the defensive against any threat of change the future might bring. Such an attitude acts as a block to progress, because the *cultural* dimensions of human endeavors are sacrificed to a strategy of social defensiveness.

Until the social sciences are willing to revise their image of man as the passive object of social forces, they will not be able to make a positive contribution, in other words one that is pragmatic and feasible, towards helping people orientate themselves in the technological and scientific process of evolution. Instead of cursing technology as the devil's work, the uniqueness and distinctiveness of the human being, his overall superiority when compared to no matter what kind of machine aided performance, should be acknowledged and made the starting point for a differentiated strategy of social action. The fact should also be faced that the modern information and communication technology, with its broad range of individual options, might very well be able to lead the way in overcoming old economic and social "role games."

The social sciences should have less trouble in granting this acknowledgement, as computers are designed to be increasingly user friendly and working with them becomes more interesting and variable in terms of logics and semantics. Unless I am completely mistaken, the humanities are on the verge of a renaissance of their cultural influence. This is due to the enormous growth in the importance of language clarity and the focus on concepts which are emerging as the most important intellectual tools of the future, especially in the computer science field. Until now, however, there has been more awareness of this fact in the technical field than in the social sciences and the humanities. The latter often feel as if society and politics have given them the cold shoulder, whereas, in truth, there is an urgent need for their participation.

In all fairness, however, one must point out that some of the popular technical reading material now flooding the market is not exactly suited to allaying people's fears.

In a well meant effort to inform readers about the undreamt – of potential of existing and future communication and information services, a downright terrifying barrage of technical terms have been let loose: viewdata, compact disc, personal computer, CAD-, CAM-, CAE-, CNC-, controlled machines, narrow and broad band transmission media, copper co-axial and glass fibre cables, satellite broadcasting, sensory technique rain down on the public and produce at first a feeling of helplessness, and put people on the defensive. This feeling is increased when the enumeration of the new devices is spoken of in the same breath with an "apocalyptic vision" of a technological society, with its range of electronic mailboxes and fully automated factories. Half the nation gets the feeling that they soon will become superfluous.

In particular, it is extremely difficult to calculate the rate at which the different innovations in information technology will become widespread. It is certain, however, that the rate at which the various techniques will find their way into production, organization and communication will vary widely. In some areas the rate of expansion has already slowed down, for example in the commercial use of large-frame computers, whereas the increase in the use of industrial robots will probably continue for some time.

This alone, however, does not tell us much about the extent and quality of change that will be involved in the different fields of work. In most fields, the new technological equipment will be incorporated into the framework of conventional working methods, and only gradually develop its own variations in use. Generally characteristic for this development will, in all probability, be the increase in the percentage of work dealing with information. This can happen within a relatively short period of time, as was the case in the printing industry (where the large printing houses, by the way, now employ more people than before the "phototypesetting revolution"). But it can also stretch over an indefinite period of time, as in the case of the teaching profession, whose sphere of responsibility is on the increase, rather than on the wane as part of the growth in the new technology.

An analysis of economic history has shown that, in the past, new technologies have taken about a quarter of a century before becoming fairly widespread. If the feeling nowadays is one of being overrun by an avalanche of micro-electronics, we should not lose sight of the fact that industrial microchips have now been around for more than fifteen years; we have already passed the halfway mark in the expansion phase. In contrast to earlier technologies, however, micro-electronics have been expanding exponentially, even during the introductory phase. The result is an increased dynamism, especially in the area of numerical manipulation.

Most of the machines and devices that are playing a significant part in changing the work world and technical communication have not been the result, however, of micro-electronics alone. Characteristic is the use of a combination of a number of different information technologies. The rate in the development of the different techniques varies widely, however (for example, with sensorics or in the digitalization of telecommunication networks), so that till now rapid progress has been made in the automated manufacturing of some mass product goods, but one cannot talk about technological changes being made by leaps and bounds in all areas of life at the same rate. We still have some time to get used to the changes and to adjust our private and work life, step by step, accordingly.

The time that we have is not unlimited, however; the range in the application of the new technologies is increasing daily, in connection with each other as well as in combination with conventional techniques. Both society and the individual must be willing and able, therefore, to learn to adapt themselves. The willingness to be flexible has to be motivated, supported, and made a matter of public interest. Those who refuse to participate in this development are standing in the way of progress, just as those who get carried away in their enthusiasm with visions of a technocratic utopia and lose touch with the realities of time and human nature are no help either. Black and white thinking almost always makes things worse, particularly when matters concern the future.

Chapter 7
Tool Age of the Future

Interrelated Systems

In trying to discover the structural elements of the information society, there does not seem much sense, after all that we have said, in simply concentrating on innovations and pitting man and machine against each other in a performance competition. The use of a number of interrelated technologies, which all contribute either directly or indirectly to an increase in the volume of information and communication flow, and the combination of interrelated systems based on a division of labor are much more characteristic features. The high degree of organization involved in a comprehensive and almost simultaneous exchange of information seems to me an important distinguishing mark between the old, and the new economic and social units communicating with each other by means of networks.

Thus, the information society is basically nothing more than a further step in the direction of the cultural evolution, which has joined people together in ever more complex and sophisticated interdependent group formations. This implies that we are still at the beginning of a chain of cause and effect, starting with the first ventures in system technology and including changes in national economic and social patterns, eventually leading to a transformation in the international division of labor. We are still, so to speak, in the tool age of the future.

There are a number of reasons for this estimation of things which, in their turn, imply certain assumptions for the future course of events.

We have already arrived at the point where the terms information and communication can no longer be differentiated. The mere storage of knowledge and aptitude would be worthless, and not justify the necessary time, money, and energy involved, if the type of access, application and transfer could not at the same time be organized along well-considered and effective lines.

In the industrial society, as we know it, information and communication are organized, to a large extent, in systems that are independent of each other. There are a host of different transmission media – ranging from spoken language over written messages to freight transportation. They all need their own infrastructure and their respective time/cost factors rise in the degree of the distance to be overcome.

Most of the new information and communication technologies are also still at a stage where they are organized along the separate branches of industry (office technology, production orientated processes of engineering, design techniques, communication techniques, and techniques for the home). Interfacing for the input, processing, and transfer of information between these different branches, however, is already compatible. This means that the transition from "outside" to "inside," and the other way round, can be accomplished within an interrelated system without costing much, as long as the corresponding technical infrastructures are adequately available.

Limits imposed by the factors of time and space are thus being done away with to a large extent, although they were, of course, often important in preserving local features and peculiarities. The flow of information is becoming increasingly internationalized (by means of world wide links between large data centers, satellite chains and cable network systems), dematerialized (not just the use of paper, but also freight haulage, means of transportation, and traffic can be considerably reduced), and collateralized (important information can be retrieved simultaneously and at any time).

This calls, in the long term, for a fundamental alteration in the structure of our national economy which is, with its large volume of foreign export closely connecting it to world trade, absolutely dependent on an optimal access to information for its survival. Tomorrow's economic rallying call will be "we have to have the information" – there is no justification for businesses not to link up with the standard systems of the new information infrastructure. Accordingly, different levels will be formed which will be in close communication with one another: the establishment of new businesses which are conceived, right from the start, along the lines of cooperation with existing research institutions and the manufacturing of hightech products based on a division of labor; craft and trade forms which see themselves paramountly as service industries with advisory functions, and which share in common up-to-date data and telecommunication devices; work units with a large degree of autonomy which, with the aid of available development, production, sales, and marketing data, are able to produce "finished" product and service packages; branches of industry which maintain joint information networks for keeping tabs on the market, for evaluating the results of pure research or for linking up with mainframe computers.

"Thinking in terms of systems and networks" – as a German business journal puts it in describing the profile and new standards for managers – will not remain limited to the realm of business. It will also be a fundamental and necessary criterion for expedient political action as the interrelationship between the different areas of life, environment, national economy, information infrastructure, social security and cultural development, has to be taken into consideration to a much greater degree than in the past. Statewide and national information systems are therefore imperative; a politically effective use of this volume of data requires the type of employee that would resemble the staff of a well organized business.

The interlinking system of science, production, and services will also lead to changes in the division of labor at the level of the international community. The leading industrial nations will strive above all to extend their dominance in the area of research. License agreements and technological trade, to a certain extent, will take the place of production output which will tend to drift to nations with lower wage structures. On the other hand, the use of machines to full capacity and automation should reduce the percentage of wage costs and lead in a limited degree to an emphasis on production. Overall, though, the importance of the manufacturing industry will decline, while at the same time becoming a necessary factor in the economic and occupational development for the expanding populations of the third world. The role of the industrial nations will be to provide developing countries with the know-how for overcoming the early stage of industrialization, with its dangers for the environment and exploitation of natural resources, as quickly as possible or, ideally, to skip this phase entirely.

Trade patterns will change accordingly: among the highly industrialized nations there will be a large increase in the volume of exchange of hightech products, information technology hardware and software, services, environmental protection measures, medical devices, patents and blueprints; the developing nations will expand their export of half-finished and finished products among each other and in trade with industrial countries. There will be a broadly diversified exchange of goods between regions, which will be subdivided along "rungs in the ladder," rather than in blocks. These goods will include computer, laser, and space technology, biological and marine processes for a number of newly industrialized nations, as well as classical investment product techniques, environmental and manager know-how for the majority of developing countries, and nutrition and training programms for the least developed nations. Protecting the latter from trying to compensate for their industrial shortcomings by the ruinous export of valuable raw materials, will be one of the most important tasks of international organizations (not just the United Nations and the World Bank but also, and in particular, the European Common Market) and a measure of north-south solidarity.

Whether these measures, as we have outlined them, will in fact be carried out depends, of course, on a number of unpredictable factors. The most important and crucial question for us is how quickly and thoroughly Europe will be able to shake off its old structures which, by now, have almost turned the industrial giants of the first generation into a Gulliver in chains. Despite a clear signal that things are getting under way, especially in the younger generation, the question is still open as to whether the economic and technological changes will be generally accepted or not. Spiraling unemployment, which in reality is not a spiral but rather a conglomerate of various causes and manifestations, still serves some as a cheap alibi for a moratorium.

Nonetheless, this brief sketch of the most important economic and technological factors can be regarded as valid. The obvious advantages of the system, the so far relatively limited degree of expansion, the positive effect on economic growth that can be expected over the long term – by the year 2000 about forty per cent of the world wide industrial production could be booked to the account of the information technology –, and the sharp calculations on the part of millions of business entrepreneurs are creating an economic momentum whose direction is clearly indicated.

Chapter 8
The Rapid Growth in the Range of Applications

Accepting the New Technology

The degree and intensity of an integrated system depends heavily on the level of modern communication technology and the reactions of private and business users. In this respect, there is also often the tendency to oversimplify in wanting to subject a network of telecommunications and computer technology to every area of life, without making allowance for any practical distinctions.

According to this stereotyped view of things, the offspring of the information age live alone from and for the "communication socket." Hooked up to an integrated digital wide-band network, the computer child communicates with the outside world like the astronauts on their first flight into space – alone and on the leash. Viewdata is old hat to him/her, and if he/she needs to order anything it is done electronically as a matter of course. Cash money is an anachronism since everything is booked automatically from the bank anyway and the balance can be displayed on the computer screen at any time. If anyone wants to buy a house or enlist the services of a marriage bureau, they simply look over the three dimensional colored listings from the comfort of an armchair at home. If you want to wish grandma happy birthday, you just dial your picture phone and give her a call. The days when one still wrote postcards or letters can only be vaguely recalled by a few oldtimers. The electronic mailbox sorts out all incoming messages, immediately discarding non-essential material and displaying important items in condensed form on the screen. The other way round, one's own computer is programmed with text elements for miscellaneous correspondence so that congratulations or condolences can be transmitted by means of the telecommunications network by the touch of a button to the receiving computer. Since a terminal is likewise built right into one's work station, one hardly ever has to get up once one sits down to work. Income tax returns are transmitted electronically to the Internal Revenue Service as well. Computer games, or the slightly boring consumption of the twentyfour different TV programs aired, make up the evening's relaxation; if one feels like expressing an opinion, an input channel with direct access to the television studio is available. This stereotyped synopsis of a day in the life of the computerized creature of the future clearly shows just how farfetched and distorted such misrepresentations can be; the human being is turned into a slave of software, part of a giant information package.

The sooner we take leave from this kind of nonsense, the better able we will be to meet the concrete challenge facing us. We need to carefully weigh the technological developments, their economic feasibility (determined for the most part by prices and fees), individual tolerance, social and cultural consequences and the subsequent need for suitable laws and regulations.

The initial euphoria at the rapid pace in the development of electronic services has given way, in the meantime, to a more rational and subtly differentiated point of view. We are coming to realize that for some time to come there will, in all probability, be two entirely separate markets with quite different prospects, one for professional users and one for a totally unpredictable mass market. After the boom in business communication did not quite live up to everyone's boundless expectations, there is now a tendency to be more reserved in evaluating the new private communication market. The "magical" boundary line of one million viewdata terminals in operation now seems as if it will not be reached until the end of this decade. The future of the picture phone is also uncertain in view of the fact that participants in a current pilot project of the German Federal Post Office are having trouble in using the new instrument regularly. Video conferences, although advantageous in terms of organization, have also proved to be psychologically inhibiting for people. Text, graphics, and data transfer services will therefore probably be limited to office and administrative work in the near future.

On the other hand, one should take care not to throw out the baby with the bathwater as – in line with the usual pattern of mood swings in public opinion – some slightly frustrated computer salespeople are already doing. The stagnation in the number of facsimile service, electronic correspondence, and data transfer package subscribers does not mean anything more or less, than that there is still simply a great deal of uncertainty as to the performance, price development, and fee charges in these areas.

As far as doubts concerning performance and price go, such hesitation is typical for the introductary phase of most technological innovations; as the case of the video recorders illustrates, however, a change of heart on the part of consumers can occur quite rapidly. A certain degree of uncertainty concerning the fee policy of the German Federal Post Office is to be expected in the early stages of converting to wide-band communication and digital data transfer, but a greater degree of clarity should emerge as things become more concrete. It would be fatal, therefore, to follow a course of development in which each party concerned waits to see what the other is up to – for the industry to sit back and wait for the Post Office to get things off the ground or, the other way around, for the Post Office to hold off until business makes clear demands – and nothing happens at all.

It is absolutely necessary to establish a powerful, integrated service network (ISDN), with a step-by-step realization of digital technology and the wideband cabling required for the transmission of animated images. We are dealing here with our communication and infrastructure lifeline of the future.

For the time being, however, most private households will not be affected radically by the changes. The average consumer can expect a more convenient telephone service, gradually more attractive viewdata offerings meant for frequent use, and a broader range of television programs, including foreign or private network channels, but not much more. Several decades will pass before dealing with multi-functional terminals becomes a matter of general and everyday routine, as depicted in the horror visions of an isolated and computerized society of the future. The reason for this is not just that it will take about thirty years for a communication network, with the capacity to meet the requirements of the wide variety of electronic services in question, to be universally and firmly established; more important: no one today can say just when people will start to feel the indispensable advantages of being logged on to computer information and service networks via their home computer, so that they would just as soon do without such services as they would relinquish their radio and TV sets today.

It took almost half a century before the radio found its way into practically every German household. Even if the new form of individual communication takes ten or even twenty years less time, this comparison alone shows how many other economic, social, and cultural changes have to be expected in this timespan. It is therefore presumptuous, and shows a lack of imagination, to limit one's projections to technological developments and to try and fix human behaviour at the status quo of 1986.

People are willing to adopt technology into their homes to the extent, that they are convinced that it will increase their quality of living; everything that is being written and said today about people's willingness or unwillingness to accept computers, is therefore generally just so much rubbish. The heated debate in the late seventies with regard to cable TV in Germany can almost be taken as a classical case of this kind of thing. There were moratorium decisions, party resolutions, and the Chancellor felt called on to point out the benefits of TV free evenings. In the meantime, the video recorder was quietly but inexorably on the march into the nation's living rooms and within five years was able to do over one and a half billion dollars worth of business, and could be found in fourteen per cent of the German households. The chance of arranging one's own TV program was obviously much more appealing than tuning into a political debate about the pros and contras of a media form that was not even on the market yet.

We can expect the demand for integrated telecommunication networks to grow much more rapidly in those regions of West Germany with a well-developed industry and business community than in the area of private households. The system offers a wide range of advantages to large corporations, not only for maintaining communication within the company as well as between the different branch offices, but also for staying in close, direct, and as extensive as possible contact with the outside world. This calls not only for the latest developments in the technology of text and data transmission, but also for the establishment of logistics for the transmission of pictures, graphics, and language geared to specific businesses, which would make being hooked up to glass fibre networks economically worthwhile.

The next step requires the permanent establishment of electronic information and service relationships between large manufacturers or service industries and their most important suppliers, clients, and independent marketing organizations. Long before the housewife can go "TV shopping" at the corner grocery store (if she ever does), a considerable portion of the business supply needs, collection, customer service and training will be dealt with with the aid of telecommunications and the computer.

This will give large corporations an edge in terms of time and costs saved, and in turn will force small and medium sized businesses to establish similar infrastructures on a local scale. Central units, such as a host computer, view-data service packages, or express picture phone service could be held in common. A wide-band network must therefore also be made available for medium sized businesses, including those in non-urban areas. The growth in the establishment of highly specialized businesses, new freelance consultant services, and spin-off companies also urgently calls for a powerful communication system outside the large manufacturing and service industry centers.

There will also be differences in the rate of development and regional impact of the information technology, and as a result, differences in terms of time, transportation, and material saved. These differences, however, should not be allowed to get out of hand and distort all sense of competition. The private sphere of individuals, in any event if they so want, will continue to remain untouched by any overtaxing or excessive demands of the new technology. On the job people will have to make adjustments, but as a rule they will be given time to adapt themselves. The range in the use of the new technology is spreading rapidly but not frenziedly.

The citizens of West Germany should therefore be able to come to terms with the new economic and technological structures and still uphold their sense of values. They should be able to see the positive aspects of the changes and to develop individual strategies for avoiding and resisting one-sided and excessive demands. This should be all the easier as almost all the new technologies deserve the rating "environmentally beneficial."

Chapter 9

Détente for the Social Environment Conflict?

Integrated Technology

A radical shift in a sense of values has taken place in Germany during the last few years, as is well known. People have rediscovered a sense of themselves as an integral part of their natural and social environment. The threat of a loss of identity through the wilful destruction of traditional values was clearly felt and averted by large-scale efforts towards their preservation. Historical town centers were able to win back their old splendor, village wells were put in working order again, a sense of care and attention for nature set new guidelines for landscape planning. Museums, the theater, and art galleries are drawing crowds in undreamt-of numbers. Poetry in the dialect and local history research have become standard features in the study of literature and history. Church congresses have become manifestations of an unconditional religious faith which is only reluctantly able to accept the need for political compromises.

The incentive of being able to buy more material goods has lost some of its power of stimulation. One should be careful here, though, of idealizing all too much: Many high quality consumer articles, to which the postwar generation directed all their mind and energy, are simply no longer of importance because they have now become taken for granted by the younger generation which grew up with them. Electronic consumer goods and vacations, however, are still considered highly desirable and a matter of status. There is even the contemptuous slogan from the USA: the long hairs of the eighties are dollars.

All the same, the importance of pure consumption has decreased considerably, giving way to a mixture of material and immaterial desires, whereby the "being" can very well take precedence over the "having" in the case of a conflict. This is especially the case when the being seems to be fundamentally concerned – for example in the area of environment and health – and what one has to give up in the way of material goods is of only marginal utility and does not affect one's standard of living. The acid test has not been made yet, though, and one would probably be surprised at how much "materialism" would come to light, if a fundamental decision between Rousseau and one's bank account had to be made, even by those allegedly fed up with civilized society.

Religious, traditional, or alternative trends also deserve special political consideration. They reflect different, but at the same time similar modes of behaviour, in that they function as a counter to the current mood of rationality. Our efficiency and success orientated industrial culture, with its tendency to schematically outline and arrange things in sequence, has obviously reached a certain stage of perfection in which the individual feels compelled to offer resistance – an act of self-assertion of post modern irrationality. The term irrationality here is not used in a derogatory sense of the word; on the contrary, it implies that after decades of cultural reason resulting from the emphasis in our civilized society on mercantilism and progress, compensation is being made for the neglect of the important emotional aspects in our cultural life.

What has begun as the open or concealed protest against the "cold" functionality of a highly differentiated industrial society, could very well help us in having less trouble than expected in coming to terms with the information society. I would like to elaborate on this theme; it may appear somewhat farfetched, however, it is no more improbable than the widely accepted theory of a disorientated capitulation to the banality of a culturally and socially deprived "computer dictatorship."

Restoring and protecting towns, villages, and local customs against attempts at intervention by industrial changes is, for example, a matter of priority for those concerned and active in the protection of historical monuments and the cultivation of traditions. Considering the rampant expansion of many manufacturing centers, there was all reason for concern. Characteristic for new industries, however, with the advances being made in microelectronics, is not only that products are continually getting smaller, but less room is needed for their manufacture. Towering smokestacks, noise, and processes involving the emission of harmful substances are not features of a technology which functions best in dust free rooms. When dealing with ambitious plans for miniaturisation, one is much more likely to breath in the air of a research laboratory than be exposed to the din of press hammers.

What are our conclusions? The cultural environment conflict between the work world, home life, and leisure time has begun to ease up. Most of the new technologies can be integrated into existing building structures and do not require separate industrial complexes which stick out like a sore thumb and are at variance with the historic landscape surrounding them.

The current structural change is much less conspicuous than the early phase of industrialization. It is only apparent behind the closed doors of existing constructions in the area of production and investment goods; the changes being made in the area of communication are completely hidden from view since they are taking place underground and in space. Thus, the familiar landscape of our civilized society remains for most part untouched, there is even the chance of partially regenerating things by the sensible use of old industrial wasteland.

The potential advantages of the new technologies in the field of urban planning do not seem at all adequately appreciated in my opinion. Policy regarding industrial landscape planning is fifteen or twenty years behind the times in many cities and communities in erecting industrial parks on the outskirts of the downtown area. City planners continue to think in dimensions of area and development along the lines of outdated factory works (gigantic machine parks, large warehouse areas, separate administrative buildings). The need for such complexes will decline drastically, however. This is actually already the case, but the fall-off has been obscured somewhat by the construction of new supermarket and warehouse centers; in this respect, however, the point of saturation has also been arrived at. Precious natural landscape reserves have been destroyed in many places by pointless industrial building while available city real estate, in both residential and industrial areas, has been either inefficiently made use of or neglected altogether. In a decent condition, such property could be made excellent use of by small or new technologically orientated businesses.

A half a century after the "Charter of Athens," it appears worthwhile to me to make a new attempt at establishing city planning guidelines for an information society. A select group of architects and city planners from all over the world met in Athens in 1933 to work out a comprehensive set of regulations to bring an end to the "chaos of the cities." Inspired primarily by Le Corbusier, the group called for an organized coordination between home, work, relaxation, and exercise, which indicated the right direction to be taken and in many details is still valid today. At that time, though, the avantgarde of architecture had to deal with a type of technology which was too expansive and voluminous to be able to be reduced to proportions which did justice to "human dimensions." Changes are already taking place today, and the possibilities for tomorrow are even much greater. It will take generations before badly planned cities can be fundamentally changed. Segmentation in city planning is obsolete and must be replaced by a new, integrated form of architecture dedicated to making optimal use of the structural changes in the area of service, production, and communication.

Accordingly, the transition to the information society should take place without affecting the external framework of our lives to any great degree. This is what makes it easy for those interested in preserving tradition to accept the new technology. Even those who are sceptical concerning the social and individual consequences of the changes have to admit that the possibility of being able to identify with one's familiar environment will be positively affected. Under certain circumstances, this could even lead to less job mobility – a factor contributing to today's unemployment problem. As long as the predominant structure is one of an industrial society, however, in which people have to travel long distances to work and more jobs are eliminated by rationalization measures than can be created by the new industries and forms of service, there will be problems. The more decentralized businesses and jobs become, in line with the new information technology networks, however, the less bearing mobility will have on the employment market. The marked capacity for integration of many of the new technologies should ,in any event, have a positive influence on our historical culture, and in the end outweigh any eventual "contra-productive" side-effects in the transitional period.

The "clean" and space saving information technology could be highly instrumental in acting as a counterweight to the environmental conflict, not just culturally but also in terms of economics. With the industrial restructuring just beginning to get underway, there is already a drastic fall-off in the use of new building land in the Federal Republic of Germany. This is certainly due in part to the enormous growth in the concern for the environment and the more stringent political and legal guidelines for dealing with public building projects. Such measures could not be enforced, however, if it did not also make sense economically to cut back on the spread of new industrial areas and to build fewer roads. People want an intact nature but they do not want to be poor and hungry in it; before things come to such a pass people will stand up and assert themselves.

Against the background of this conflict of aims, the industrial society, in its conventional form, is incapable of really reconciling this disparity between economy and ecology. Reconciliation implies the voluntary and nonviolent congruence of originally different points of view which always involves sacrificing one's earlier claims to a certain extent. As long as the "smokestack factories" form the economic backbone of our nation, however, we cannot renounce these ecologically harmful industries without being able to offer an alternative since way too many lives would be put at stake. It is also not a question of minimizing or making up for the harm in some way. In both cases it would be like trying to atone for batting someone over the head by offering to bandage him up afterwards.

The only procedure which can provide an authentic protection for the environment is one that does not allow any damage worth mentioning to occur in the first place. Almost all such procedures that we know of are the re-

sult of the combination of various information technologies in conjunction with developments in cybernetics, biotechnology, and solid-state physics. These basic technologies alone make the following possible:
- the protection of natural resource land as less building land is required;
- no demand on natural resource land for purposes of communication or transportation;
- less use of resources and energy;
- the technical substitution of products whose manufacture involves the emission of large quantities of harmful substances into the environment;
- measurement and control procedures for a permanent control of the environment which are a built-in feature of the industrial manufacturing process;
- the adaptation of continually more production concepts in a system of regulated cycles which resembles the natural recycling process;
- an increase in production which allows for new service investments, especially in the environmental field.

The basic characteristic of these technologies, which allows for a synthesis between economy and ecology, is a measure of their achievement: The larger the quotient is between the capacity for immaterial information processing and the necessary material investment, the more successful they are. The trend is therefore in the direction of immaterialization, towards the absolute limits of physics. This is something fundamentally different from the criteria of success in the steel and iron industry which is based on multiplying material power and attributes.

Computer simulation, in particular, vividly demonstrates how information can replace material. Prototypes and test procedures are becoming increasingly superfluous in the automobile, aircraft, and mechanical engineering industries because the property and behaviour of the materials, tested under a wide variety of circumstances, can be calculated and depicted graphically. The three-dimensional sequence of movements on the computer screen is an exact simulation of reality and can be reproduced any number of times or changed to fit new measurements. Once the computer has calculated the optimal construction data and displayed the design of the machine or systems part on the screen, production can begin – without any further trial runs. Experiments, which are costly in terms of time, material, and money, can be done away with. The greater the calculating powers of the computer are, the more possibilities there are for simulation. The most modern computers can calculate to one hundred per cent the "environmental behaviour" of even such complex systems as nuclear reactors, aircraft, and molecules. This means not only a higher standard of safety for man and nature, but new medical insights can also be won in this way.

The consistent use of the new technologies serves to alleviate the conflict between the economy and environment in our industrial society – this is demonstrated, not least of all, by the consumption of energy. Between 1960 and 1973, in the Federal Republic of Germany, the industrial consumption of energy and the gross national product increased in proportion to one another. Since then there has been a drastic decrease in the correlation between economic growth and the increase in the use of energy, which still continues to this day. Till 1982 the domestic product increased at an average annual rate of 2.1 per cent, while in the same period of time the industrial consumption of energy only increased at the average rate of 0.2 per cent. Almost eight per cent less primary energy was consumed in Baden-Württemberg in 1983 than in 1979, although there was a 3.2 per cent increase in economic growth. The will to economize alone would not have made this possible. It brings rather to light the "multiple savings effect" of micro-electronics: Many modern industrial products, for one, only need small amounts of electricity, secondly, micro-electronic measurement, control, and regulation technologies make a more rational use of energy possible, and lastly the computers and data processing systems within the power stations themselves guarantee the precise distribution of energy according to need.

Technological progress is continually doing with less and less environmental elbowroom. Not only that, it is beginning to slowly but surely revitalize nature. This certainly has to be a powerful incentive for those who are concerned about the growing loss of trust in a culture dominated by industry to give the information society, with its more flexible and ecologically beneficial technologies, a try.

Yet, incorporated within the outwardly inconspicuousness of many modern production and procedure technologies is a political problem which should not be underestimated. Since their cultural and ecological advantage is exactly that they do not affect the substance of things, it is not particularly apparent from the outside. On the other hand, the economic problems involved in the transitional period and the potential misuse in the area of personal data, as long as clear-cut legal regulations do not exist, are obvious. In other words, the short term and hypothetical dangers can be more clearly perceived than the much more substantial advantages, to which not only ecological but also economic growth can be counted.

This makes a reasonable and moderate technology policy vulnerable to attack; and the attacks indeed are not failing to materialize. Some of the very same people (now older but hardly wiser), who wanted to make a bogyman of the Federal Republic of Germany during the sixties as the governmental incarnation of lies, deception and surveillance, are still at work. At that time professors, the police, and politicians symbolized the "repressive system," today diabolical powers are at large who want to coerce glassy eyed citizens with the help of a gigantic online data storage systems.

So far, no one has been paying much attention to this kind of polemics, not even at the universities. Today's students not only grew up with consumer and recreational electronics, they have even, to a certain degree, consciously taken up computer technology as a distinguishing characteristic of their generation, as their form of "protest" against the older generation. You can not get very far using technological arguments in trying to inspire a generation used to technology to reject the government; the young people of today are much more apt to get a kick out of proving their technological versatility by playing a trick on the government, as can be seen in the popular "hacker sport."

Nonetheless, vigilance is called for to avoid the use of false arguments and excuses for bringing about the type of economic standstill we faced at the beginning of the last decade. Now, as then, those who have to suffer the consequences are not the objectionists themselves, although the beleaguered and suspect government is certainly no longer in a position to offer the shelter of as many tenured positions. Those to suffer would again be the average wage earners, who would have no other choice but to wait out the next decade with its increasing risk of unemployment as industries decline.

From a political standpoint, therefore, the positive option of the new technologies has to be kept open. This can only be done within a framework of social and economic policy which consciously promotes social, cultural, and individual currents of thought and standards of value, and makes specific use of the potential for progress of the new technology to win more in the way of freedom, self-determination, solidarity, and responsibility for the future.

A politically credible compromise between the "technocrats," on the one hand, and the "mythocrats," on the other, as Edzard Reuter recently described them, has to be found. A policy of renewal must indeed be on its guard against any kind of bias, since people, as well as technology, are required in order to achieve its aims. The technocrats must not be allowed to disregard the social risks and personal fears involved, whether they deal with work, certain standards of training or personal data protection. Under certain circumstances, making use of the technology will have to wait until the legal and political questions can be satisfactorily resolved. Baden-Württemberg has got a political head-start on the rest of the Federal Republic of Germany with its State Media Law, but it was one of the last bills to be introduced into the parliamentary debate. Some people were surprised at this, although there is nothing surprising about it at all. The subject matter is extremely difficult, and we have had very little experience with it. Countless rounds of talks with experts, organizations, social and political associations were held. The legal and constitutional aspects, data protection, competition and youth protection rights were meticulously examined. Every conceivable economic and technological constellation was repeatedly gone into in an attempt to determine its compatability with the constitutional guarantee of freedom of speech and liberty. Our timetable was not dictated by technological or commercial interests, but rather took into account matters of political priority. Furthermore, the law was restricted, from the start, to a period of five years, in order to have a free hand to make any necessary revisions arising from new developments in the interim.

On the other hand, the "mythocrats," with their fatal tendency of viewing technology and culture, progress and the individual as being permanently at odds, cannot be allowed to get the upper hand. A technologically orientated economic policy therefore has to go hand in hand with an active and progressive cultural policy by the government, which makes use of every opportunity to do justice to the increasing interest of the individual in the fine arts and historical subject matters. In principle, it would be underestimating the social consequences of the technological change in structure, if the government tryed to limit itself to promoting scientific research, economic innovations and a fairer distribution of social benefits. The museum caching of industrial culture, as Hermann Lübbe has dubbed the widespread interest in the preservation of museums and the environment, clearly indicates a basic need of the human being to compensate for the often overwhelming claims and demands on his/her rational person, in the area of public life, by creative and irrational activity in one's private life. The basic pattern of a highly integrated information network, which more and more people will be confronted with on the job, will therefore have just as little influence on recreational life as the mechanical sequence of work in the industrial world did. People will not let the opportunity pass them by of communicating with one another, of lending a hand themselves, of gathering their own experience, in short; to experience the world with the help of their own natural senses. They will be more than willing to relinquish certain routine jobs of everyday life which lack any sense of self-development to the charge of technological slaves, just as they willingly bought washing machines and dishwashers. The need for social and cultural commitments, in which one can and must be personally involved, however, will grow enormously.

The everyday cultural side of the information society should be much more like a colorful patchwork quilt than a gray cement wall. Different job and private preferences in values will develop alongside of one another, if not always successfully, and with strong deviations between different age groups and stratums of society. One characteristic will be a high degree of individuality, however, which will give the lie to the distorted picture of a standardized computer society of human robots.

Chapter 10
A New Wave of Independence

A Flexible Work World

Work belongs to the everyday life of an industrial society. A crisis in the work world is also society's crisis. The information society will also be shaped by the work world; between the economic sector and individual fields of activity, however, there will be a large degree of movement. The proportion of service related jobs (and in turn the share of information and communication associated jobs within this sector) of the total employment market will increase. This is one of the central features of the structural change and we will be going into it in detail; the same holds for the changes in the disposition of time and social patterns, which are part of the same coin.

In the secondary field of production one fears the worst in connection with the new technology. Disadvantages, not only for the quantity, but also for the quality of work, are often expected: fewer and more monotonous jobs.

It is certainly true that factories are being put under a lot of pressure to automate, especially with regard to assembly line work and the processing of workpieces. The trend towards machines replacing human workers cannot be overseen. Till now, this fact, however, has only played a secondary role in the overall cause of unemployment. The accumulative effect of demographic, educational policy, and cost-intensive factors outweigh the technological components by a long shot.

In addition, there is what one can call the "iron law of structural conformity": in making adaptations to changes very early one runs less risk of ending up with a negative effect on the employment market, even in the transitional phase. The new products and procedures are able to enjoy an edge over the competition and lead to a greater share of market, expansion of production and the hiring of new employees. Conversely: the slower one is – in comparison to the international competition – in adapting to structural change, the more one is forced by the question of cost to undertake short-term rationalization measures. In other words, it takes much longer for product innovations to break into markets already occupied by others and to establish new production and job capacities. We end up paying double and threefold with unemployment, one could say, with every year we procrastinate.

Germany's entry into the age of micro-electronics was delayed by several years. However, this cannot be placed entirely at the doorstep of the business community. The politics of the seventies held things back and were a hindrance, rather than a help in promoting change and getting in step with high technology. A policy mix of supporting the economy and technology, along with certain tax and employment market measures, can help to shorten, but not entirely do away with a phase of unemployment due to the required adaptation to structural changes. The shadow of unemployment will continue to hang over us throughout this decade, as much as we would like to try and shake it off.

As far as computers and a change in the quality of work go, we have to remind ourselves that it is for the most part work involving health hazards, noise, and the emission of harmful substances which is being taken over by machines. In view of the annual statistics concerning industrial accidents and early retirement due to ill health, computers can be considered as making an important contribution to the humanization of the work world.

But also in terms of an overall picture of the workaday world, organizational forms in which people are simply cogs in the wheel are not on the rise. On the contrary, there will be automated branch systems – particularly in the production, assembly, packaging, and loading of mass-production products – which will call for more group autonomy on the job with a relatively higher degree of freedom of choice, for example, in the area of production planning, programming, design, and the drafting of customized solutions.

In the future attention must be concentrated primarily on using the large potential of flexibility of the information and communication technologies in achieving an optimal division of labor between man and machine. It is not sound tactics to simply assign people to ostensibly autonomous auxiliary tasks in the expanding network and automation of the on-the-job flow of information and material, in order to have something, so to speak, to fall back on in case the self-regulating process should fail. Although the need of being in direct contact with raw material and making articles by hand continues to be on the wane, control, supervision, and maintenance activities make up an increasing proportion of work time. At the same time, it is just these types of activities which are experiencing a new growth in importance as automation itself becomes more flexible and sophisticated and the technological areas of application, which one could describe as "semi-automated customized work," are continually expanding.

There is a wide variety in these forms of production which differ in accordance with the size of the concern and the type of product being manufactured. They include "insular" computer-controlled machines, which only carry out certain processes and require a good deal of manual support, so-called CAD/CAM systems, in which the calculation, construction and manufacture of parts are carried out by machines, and the actual "flexible" production, which also allows for the program-controlled production of a limited number of items. Especially in the latter case – the use of which has not become very widespread, as of yet, due to large investment costs involved – the enormous expenditures are only worthwhile if a considerable expansion in the diversity of production can be launched at the same time. It has to be possible, in other words, to convert plants quickly and with a minimum of expense to be able to manufacture small amounts of diverse series and to carry out individual customer wishes. Only then do the investment costs add up. In limiting oneself otherwise to just a few, unalterable finished products, there is a good chance that the amortization period of the machines will last longer than the lifespan of the products themselves.

What is our conclusion? Regarding automation, it is not plants manufacturing mass-products in large series so much as middle-sized businesses which are absolutely dependent on technological flexibility. The ability to quickly adapt to new market trends is impossible without employees who can think and decide autonomously. "Surveillance drill," simply being able to react to technical alarm signals, cannot, by any means, be judged as an adequate qualification for this type of work. This explains why one has been able to observe a trend in the last few years of cutting back somewhat on the functional division of labor with an increase in the demand for skilled workers who are capable of "workshop programming" for automated production work.

A comparative study of the International Institute for Management and Administration by the Science Research Center in Berlin and Henly Management College came up with some quite interesting conclusions regarding their experience with computer-controlled CNC machines in Great Britain and the Federal Republic of Germany: According to the study, there is a trend in both countries towards the manufacturing of small and frequently changing series. Such series can only be produced with computer programmed machines, and not numerically controlled ones – as with the NC-technology of the seventies. The possibility of carrying out flexible machine programming with the help of CNC-technology opens up a new, distinctly middle-sized field of operation for automation. This development has led in Great Britain, which has a relatively small core of skilled workers, to a further division of labor between production management (writing or adapting machine programs) and production engineering (maintaining and supervising the machines). British master craftsmen or foremen rarely do workshop programming work. In the Federal Republic of Germany, on the other hand, there is a much larger proportion of skilled workers, who can independently carry out the programming, setting-up, and maintenance of CNC machines. In contrast to the inflexible NC production method, this has already resulted in better qualified skilled workers and acted as a check to further unemployment.

About 98 per cent of the commercial enterprises in the Federal Republic of Germany can be considered as basically middle-class in character. They employ just about two thirds of the working population and train four fifths of the apprentices. Their strength lies in customer orientated work to order, along with a well-developed capacity for adaptability. These advantages can be consolidated by utilizing flexible automation technologies. This calls for the same qualified and independent workforce as in the past. There is just one, however, very important difference: Self-reliance must be shown more in understanding the process of manufacturing, rather than in work done by hand. This process can be learned, however, when both the business community and the unions recognize and support its priority.

Can we make the same kind of optimistic predictions for large concerns and office and administrative work? I believe we can, although a somewhat different type of strategy is called for. Without a doubt, the idea of using people as system components in a close-meshed network of screen, data, and control stations is not so remote, and essentially entails nothing more than a new information technology form of a division of labor along the lines of Taylor. In this case, though, the limits of being able to cope psychologically are so clearly marked, and the alternatives of a humane and efficient cooperation between people and technology so obvious, that it would be sheer stupidity to permanently resist a flexible and individual structuring of work.

There are many possibilities for doing justice to the new demands on cooperation, a number of which have already been successfully tried out. There is no need here of going into detail about the well-known examples of job enrichment and job rotation. More flexibility in organizing work time, after the protest roar on the part of some associations has faded away, will gain widespread acceptance. Making optimal use of machines is not the only, or even the most important reason, for this inevitable development. Employees themselves are increasingly coming to feel that rigid work times are an anachronism in an age in which – in the media and advertisements – independence, self-determination, and using one's own faculty of judgement are held up as the most important virtues of a citizen.

Important findings in the field of industrial medicine support this view. By the end of this decade, at least according to the calculations of the computer industry, every sixth workplace will be equipped with a visual display unit. The logical consequence of their increased use is more part-time job models and the spread of home computer work, since no one can do concentrated work for eight hours in a row at a computer screen. Estimates by American institutes predict that at least twenty per cent of employees in the United States will work at home by the turn of the century. As a report by the Congressional Budget Office stresses, an especially important role here will be played by women, who often see computer work at home as the first step in later acquiring their own small businesses.

In contrast to this, political reservations and social doubts dominate the discussion in the Federal Republic of Germany: unemployment insurance and social security are tied up to a minimum of working hours per week and health insurance benefits are dropped if one takes off from work without pay for any longer period of time. If the unions and management should not be able to arrive at a satisfactory solution to this problem, which would be infinitely preferable, it is up to the lawmakers to provide adequate protection for such forms of employment by passing appropriate industrial and social legislation. This could be done by changing the basis for calculating the minimum required working hours (for example, calculations on a quarterly instead of a weekly basis) or by guaranteeing the right to apply for social security for those employees, who, till now, have not been subject to compulsory insurance. Legal measures could also be initiated to clear aside uncertainties with regard to protection against wrongful dismissal.

This complex of legal steps that can be taken, however, should not consist of isolated meausures by any means. They should be part of a general, absolutely necessary effort, to make industrial law, both in terms of the individual and collectively, more flexible. Both employee and employer must be granted a greater degree of self-determination in organizing their work, both as regards content and duration, and in structuring wages and concrete working conditions. The new technologies incorporate an essential element of independence, beyond simply part-time or at home work, which should be taken advantage of.

Many employees, undoubtedly, would be interested in being able to choose between a more routine or high pressure job with more free time or better wages, on the one hand, and an occupation with more independence and greater demands on one's time, on the other. Others might find it advantageous not to have to punch a time-clock and be willing to accept a salary split up in the form of a basic wage and success premiums. Department staffs could consider whether to combine heavily schematized areas of competence into group orientated "integrated" work tasks and to determine themselves the ideal time schedule and form of organization.

The technology is available for organizing work along such variable lines. In no other earlier stage of technology were conditions so favourable for human autonomy in industrial and service jobs as the one we have today. What we are lacking are imagination and courage to adapt the growing technological flexibility to individual rhythms of life and work. It is more than annoying that the lawmakers, unions, and employers, in a rare demonstration of solidarity, all shy away from taking decisive action and cling to traditional patterns of institutional order. They create mutual alibis for each other, which leave the rigid structures of law, time, and organization, for the most part, untouched. Pioneer work can and must be done, above all, at the level of concrete industrial agreements. Only then can the operation and system designs, which have been paid scant attention to till now, receive the necessary practical impetus for adapting machines to fit the needs of the people working with them. This, in turn, could create a market in which the Federal Republic of Germany could take the lead. Our conclusion: There are not only technological and humane, but also concrete economic reasons for choosing a course of more flexibility in the work world.

If we bring these reflections on the possible new forms and content of work into relation with our findings regarding the conserving character of the information technology, both in terms of nature and culture, we cannot avoid the conclusion that what we have here is the chance of being able to shape everyday life, both "externally" and "internally," to be more in keeping with a human measurement of things: Large industrial complexes, which destroy the landscape and throw the ecology off-balance, are becoming more and more of an exception. While they still continue to exist as manufacturing centers for traditional products, the economic pressure to convert to automated mass-production is growing. This is the only way of coping with the pressure of competition which is reflected in price wars and product refinement. People, seen in global terms, are gradually being pushed out of the "old" industries. People are finding jobs in the branches of large companies in which special high-technology problem solving is called for, or building a new middle-class, which makes its living with flexible custom-made work and high quality service. In both cases, negative consequences for the environment are considerably diminished, while the share of self-determination in shaping one's own work rhythm is increased.

The wave of new independent businessmen/women, which one could observe in the last couple of years, is part and parcel of this line of development. To simply want to interpret this trend as a reaction to unemployment and the glut of college graduates, would be shortsighted. It reflects rather a shift in values, which makes calculated use of the new technology. The desire for autonomy in determining when and how long and what kind of work one does, is still a matter of high priority with many young people, who cannot resign themselves to the highly standardized workday world. Micro-electronics, opto-electronics, sensorics, and software engineering are the typical fields of the new technology, in which they recognize the opportunity of achieving economic independence without needing a great deal of capital. The early phases of new technologies always involve numerous problems of detail – especially in the area of measurement, control, and regulation – which are open to individual solutions. They are the best breeding ground for young businesses. Mature products, on the other hand, require above all a knowledge of production and marketing, design refinement, and service extension; they necessitate the kind of organization which inevitably curbs individual initiative.

It is by no means as if Germany simply conjured up the new technology orientated businesses out of some wonder-bag of economic tricks because they were so impressed by the examples of California's Silicon Valley or Route 128 near Boston. More and more self-employed businessmen/women with ideas for using micro-electronics also appeared on the scene in Germany during the second half of the seventies, the logical consequence of a greater freedom of the individual with regard to value conformity and technological options. They simply had a lot more trouble in gaining a foothold than their American counterparts.

Capital investors, who were willing to take risks, could not be found because German tax law is not designed to differentiate between actual and apparent risks in the form of predetermined loss allocations. The stock market was off limits, since the German stock exchange is the exclusive realm of the well-to-do. Banks demand security, preferably in the form of real estate. Special, technology orientated sponsor programs were as good as non-existent; the attitude of the federal government, at least, was indifferent or negative concerning information technology. Due to a lack of qualified personnel, the advice provided by chambers of commerce for businessmen/women left much to be desired. The universities simply disclaimed competence with regard to the matter; even the ability to assess matters from a scientific or technological standpoint often went beyond their ivory tower image of themselves, which in turn was reflected by bureaucratic reglementation of privately financed research on the part of the political administration.

Everywhere you looked: normative, social, and financial barriers – no wonder that the most active young business people and the best young up-and-coming scientists emigrated to the USA in droves, a country which attracts talent like a magnet, and is able to turn it to account in establishing its position as a world leader. Forty per cent of the basic research being done in the States today is in the hands of non-native born Americans.

Things are now gradually taking a turn for the better in Germany. New enterprise centers, predominantly in university towns, are being established, technological and commercial consultant services are becoming available in even the more outlying districts, government sponsored programs are helping to bridge the gap between the conception of an idea and its actual marketing, and the conditions for authentic venture capital, associated companies, and easier access to the stock market are improving. This development makes up in part for the injustice done to a generation, which had to see its chances at self-realization being curtailed by the short-sightedness of social politics. The same holds true for the new research programs, which are significant as an indication that there has been a change of climate, as well as on account of their actual content.

There is no denying, though, that as soon as one German sets about tackling a problem, it does not take long before two others start pointing their fingers at him in warning. Things have barely gotten underway, and we already have a fraternity of diehard doomsday economists, who feel that the government's active measures are carrying things much too far. They see technology factories as hothouses for incompetent dreamers, sponsor programs as fat pickings for unproductive career applicants, and advice centers as second-rate schools for technological hicks. In their eyes, the state is forging more fetters, intervening, basing its selection on limited knowledge, distorting the competition, and neglecting tried and tested businesses. Above all, they feel that the government is trampling on the spirit of free enterprise, which thrives best in an environment in which politicians give no thought to productivity, competition, export aids, and helping people to make a livelihood.

It is apparent that we are dealing here with a peculiar tendency of trying to prevent progress by idealizing the past. There is basically very little difference between this type of thinking, and the attempts on the part of anti-technological zealots to denounce the information society as a social wasteland. Their twisted logic dictates that our present industrial society *must*, perforce, be a treasure house of social contact and personal satisfaction, full of adventure, because the computer per definition is inhuman and isolates people. By this token, female factory workers, who perform the same mechanical operations eight hours a day, rush to pick up the children from the babysitter, run out to do some last minute shopping, to barely get home in time to make dinner, are living in a paradise of social communication. In juxtaposition to this, is the way a psychologist at a Social Democratic Party hearing described the terrible fate awaiting them as computer homeworkers, and which simply cannot be imitated: They "would lose all sense in the value of their work, recognition by others, a broadening of their social horizon, and contact with other people – at the same time they would suffer the negative consequences of being family centered (and I quote!) mothers, as well as the too restrictive, too frequent, but not very intensive contact with the children, and the lack of stimulation from outside interests which they could share with the family" (*Frankfurter Rundschau*, December 8, 1984). Factory work as a wonderland of conscious living and a prophylactic against the dangers of "family centered motherhood" – the boundary between naiveté and agitation overlap.

An economic mystification, however, is just as inappropriate and harmful as a social one. In both cases, people would rather close their eyes to the drawbacks of the old system than give the new system a chance. The shipping agent who started out with a handcart after the war, the painter who was able to set himself up in the contracting business by buying some cans of paints and a few brushes, the enthusiastic do-it-yourselfer who worked himself up from tinkering in the backyard to running a factory are all worthy of memory, but no longer adequate criteria for coping with the demands of today's business world.

A modern policy of economics can no longer ignore the following facts:

- in many of the industrial nations with which we are in competition, the government plays an active role in the development of the economy and technology;
- the part played by products and processes which require intensive research is becoming increasingly important;
- in keeping with this, capital investment costs for buying new equipment and machines are sharply on the rise;
- these developments involve risks, particularly for middle-sized businesses which are the backbone of the employment market, as product cycles replace one another in increasingly rapid succession and shifts in the market have more far-reaching consequences;
- international competition is becoming more intensive as the need for new foreign markets to compensate for the glutted domestic markets grows;
- the export of products by middle-sized companies is especially dependent on a powerful, easily available and practice orientated infrastructure of knowledge and information.

The recommendation of lowering taxes, reducing the period of time in which investment deductions can be made, and passing fewer administrative regulations, nonetheless, is still valid. They cannot be taken as cure-all remedies, though, and to deny the important role of the government in coordinating and structuring policy between the economy, science, and society, is like trying to fly a jumbo jet using the ground contact flight regulations for single-engine propeller airplanes.

Typically, it is usually those who are already well-established, who are the first and loudest to cry out against any form of government intervention. Their personal experience leads them into making the mistake of underestimating the difficulties in starting up a business nowadays, and overestimating the government's willingness to act as a control function by alloting financial aid. They cling to the notion that if political actionism would only leave business alone, it would flourish like the garden of Eden. This just is not the way things are, though: A change in structure occurs on the basis of a change in the infrastructure, and this influences the general conditions of life, work, and the economy, which, in turn, cross not only economic boundaries, but also those between countries. This development can be traced, starting with the trade routes of the ancient world, and including the ports of the Hansa towns, right up to our present day network of railroads, highways and airplane routes. The same principle applies to policy pertaining to the support of science, foreign trade, employment and education.

Strangely enough, hardly anyone objects to regional sponsor programms, in keeping with the principle of giving everyone a slice of the cake. And oddly enough, anyone is entitled to government supported retraining programms. It is just when someone chooses to be self-employed, that he is forced to master the situation all by himself. Luck in surviving the mistakes of the early years, somewhat shaken but still alive, is taken as proof for their right to exist.

The idea of a wage earner being bound to his workplace by a rigid system, dictating when and where he is to work, is becoming as outdated as the image of the selfmade man/woman, as someone who has to take the plunge and learn to swim or sink without even having the benefit of a few swimming lessons or a lifebuoy within reach. The chance of being able to set up shop on one's own, by marketing an innovative idea, is the more easily realizable the less irrelevant red-tape one has to cut through at the beginning. Financial and administrative assistance can help new businesses to get off the ground and develop autonomy, till a reasonable time elapses in which one can expect concrete results in the form of marketable products. The government could give such businesses the benefit of a doubt in the early, critical stages in which such help is most needed and hardest to obtain from private sources. Without this kind of trust, we run the risk of making technological progress the exclusive prerogative of the best equiped research laboratories – rigor mortis for the structure of business would be the result.

The market tests a new business soon enough, when it introduces its product; by then, though, it has more to show than simply a good idea and good intentions, and the chance of maintaining a hold are much better. A policy of government support for technologically orientated business can and does not expect anything more than this.

Chapter 11
The Broad Field of the Information Society

The Service Industries

The new technologies will have their greatest impact, it is generally assumed, in the large, growing area of services. Some people seem to equate the terms "information society" and "service society" as being one and the same thing. They are extrapolating a development which has already been verified, in fact, by the steady growth for the last thirty years in activities belonging to the "tertiary sector." According to a study by the OECD, in 1954 26.7 per cent of the working population in the industrial nations of the West were employed in agriculture, 34 per cent in manufacturing industries, and 39.3 per cent in the service sector. By 1977 there was a shift in the correlations: 11 per cent (agriculture), 34.5 per cent (manufacturing), and 54.5 per cent (services).

Furthermore, due to different economic structures, there are considerable differences between the individual industrial countries: In Japan and the USA, for example, trade, as an independent form of business, has traditionally played a more important role than it has in Western Europe, so that there has always been a larger proportion of people working in the tertiary as in the secondary sector in these countries. In the Federal Republic of Germany, on the other hand, the service sector did not overtake the industry until 1975, later than in almost any other industrial country. This is a clear indication to what degree our economy is still linked to structures of production and technologies, which were developed during the last century.

The Nuremberg Institute for Labor Market and Occupational Research, which is responsible for probably the most thorough investigations in this area, places the proportion of people now working in the public and private service sector in Germany at 54 per cent. However, the percentage of employed who, according to their job description, are not directly involved in the production process, is much higher, namely almost two thirds of the working population. In view of these facts, it would seem that we too, like the other highly industrialized nations, have taken a step in the direction of a service society.

What is more important and interesting for us, however, is the question as to the development of activities concerned with the production, processing, and distribution of *information*, and how much weight they carry.

The evidence clearly indicates that the rapid expansion of these "information jobs" is actually responsible for the change. Their importance in shaping the work world has grown from year to year; while about 40 per cent of the working population dealt primarily with knowledge, information, and communication in 1980, by the end of the decade at the latest, based on an annual rate of increase of one to 1.5 per cent, this figure should reach the fifty per cent mark.

It seems logical, therefore, to me to take leave once and for all from the traditional three sector model – agriculture, production, and service –, and to establish the area of information as a fourth, independent sector, which incorporates a reorganization of the traditional forms of work and economy. The dynamics of this process become especially apparent if you stop to consider things from a different point of view: What would be left of service jobs if you did not include the information professions – scientists, engineers, information engineers, bank and insurance clerks, office personnel, lawyers, teachers, advertising people, real estate agents, artists, technical designers, journalists, not to mention politicians? They would be reduced by more than half and the proportion of people working in this area would stagnate at a level which was reached twenty-five years ago; and barely one in three would be making a living in industry, with a further downward trend.

Are we already on the threshold of an information society without having really taken notice? The thought is certainly sobering. It implies that the post-industrial age would be nothing more than an aggrandizement of our own age. Will there simply be fewer industrial workers, as better machines are able to ensure an increase in the rate of productivity, and more in the way of banks, insurance companies, administrative work, lawyers, and fast-food restaurants?

To assess the matter strictly from a quantitative point of view, however, is inadequate. As long as one does not differentiate between the various forms of service and informations activities to try and determine the *qualitative* changes being made in the economic, technological, social, and cultural fields, the face of tommorrow's society will remain obscure. One can develop a series of parameters – income, shifts in age-groups, preferences in demand, work and recreational budgets, technological developments, educational requirements, cost structures, the political framework –, and on the basis of them make certain deductions as to how life in the future, in all probability, will be. In describing the "quaternary" information and communication sector, however, it is especially important to always keep the part played by *culture* in influencing the economic structure in mind.

The essential characteristics of the new information society cannot be comprehended by simply trying to calculate future output within the same old rut of worn-out economic theories. Along with the question as to what kind of services and goods will be produced, it is equally important to ask *how* they will be produced and to what extent important social and cultural values, such as interpersonal contact, self-determination, and the protection of the environment, will be affected.

Not until a new system of values has been established which can resolve the contradictions of our industrial society – economy and ecology, heteronomy and autonomy, group interests and individual concerns, the anonymity of social welfare and personal responsibility, work and recreation, culture and consumption –, should we talk about the information society as if it already existed. Establishing these standards means prolonging the transitional period. In my opinion, this is justified and necessary in order to have enough time for evolutionary changes in individual and social patterns of behaviour, which inevitably have to go through an experimental phase. Nothing would be won by rejoicing too early and trying to act as if the new form of society was already established, when we are, in fact, still living in the shadow of the old one. Currently, the information technology is continuing to make strides in changing the quality of work in the industrial sector. The changes being made here are more obvious than those in the area of services. Until, not just an increase in the percentage of people working at service jobs is registered, but also a distinction as regards content between old and new structures can be made, can one speak with conviction about a change in society.

In attempting to characterize the structure of the new services, one has to keep several aspects in mind at the same time: What fields of activity will play an especially important role? What kinds of qualifications will be required? What form of organization will prove most effective? What sort of political guiding spirit will have to be developed?

Let us begin with the field of professional "top intelligent" services, which the theoretical founders of the post-industrial society, Daniel Bell and Jean Fourastieé, have declared to be the criteria of identification per se for the socio-economic change. Personal service, in the form of servants or household help, according to Bell, was characteristic of pre-industrial or early industrial society, while business and transportation services influenced the structure of industrial society. The post-industrial phase, on the other hand, was dominated by "intelligent" services – first of all in the preeminence of scientists and engineers, and secondly in the form of new systematic guidance and control functions, which also made use of scientific methods, for example, in administrative work. Theoretical knowledge became autonomous, and by means of computer science, cybernetics, system analysis and synergetic forms of organization it was able to make strides and become a decisive factor in social planning. Innovations are no longer the product of knowledge acquired by technical or skilled work, as they were in the last century, but rather the methodical result of the planned and guided development of technology. Inventions are becoming less and less a matter of chance.

This is exactly what is happening in the elitist public and private research centers all over the industrial world. What appears to be a confusing diversity of various scientific activities is, in truth a concerted effort to penetrate the mysteries of micro and molecular structures. Within a decade the memory capacity of the few millimeter squared silicon chip has been increased by sixty-four-fold. The new generation of chips, which have room for 256,000 bits of binary information on a single component part, are now on the market but will probably not be long-lived. Highly integrated relays with a million memory units are expected to be on the market within a year and by the beginning of the next decade chips with a memory capacity of 16 million information units will probably be standard.

The technology involved in producing this future generation of chips, however, will be much different from the one employed today. The structure of silicon does not allow for a conductor track of less than a ten thousandth of a millimeter and traditional optical etching procedures are too inaccurate at a magnitude of less than a thousandth millimeter. Submicrotechnology is involved, therefore, with exploring new physical-chemical etching processes while research in the area of component parts is being devoted to testing different types of semiconducting materials.

Both lines of development require advanced knowledge of molecular technology and are therefore directly related to molecular biology, an area of science in which rapid progress is being made in the manipulation and synthetic production of micro-organisms. Some organic molecules are able to function as electric conductors and, in the area of molecular-electronics, measurements are smaller by a factor of one thousand, although processing speeds could be as much as ten thousand times as fast as with highly integrated micro-electronics. In view of these facts, the idea of a "bio-computer" is tremendously fascinating, both from a scientific and commercial standpoint. Despite the many obstacles to be overcome, the links being forged between biogenetic and micro-electronic information technologies indicate a new dimension in research which could quite possibly lead to the use of atomic and molecular structures.

Similar developments can be seen in other disciplines. New inorganic, non-metal materials are often based on compounds of silicon, the fundamental material of micro-electronics. On the basis of advances made in our knowledge of molecules and crystals, we can predict the composition of these materials and how they will react under extreme conditions with a high degree of accuracy, or they can be determined quickly and precisely with the aid of computer simulation – old-fashioned empiricism is no longer possible or necessary. Just as silicon and electronics acting in combination with one another were able to produce the computer, a compound of silicon and aluminium is in the process of replacing the classical heavy and light metals, as the basis of industrial products, with ceramic materials in an almost unlimited variety of organic and inorganic forms. The computer and the ceramic age are not separate utopias, but rather different branches on the tree of molecular and atomic knowledge which is continually sprouting new stems. Laser technology based on chemistry is just as much a part of this as the developments being made in the area of X-ray and infra-red techniques. Combinations of ceramic material and X-ray-lithography processes are being used today in the development of even more powerful chips and demonstrate the profitable fusion of different technologies.

The explosion in the interdisciplinary science of micro-structures, which interweaves biology, chemistry, physics and information technology, has had wide-reaching consequences. Branches of research, which up until a short time ago were considered as belonging to largely separate categories, are now becoming intertwined and, with the enormous improvement in instruments and refinement of procedures, mutually helping to speed up the development process. Pure research, in the narrower sense of the word, is becoming an exception to the rule because high investment costs are forcing researchers to justify their objectives in terms of their practical application. On the other hand, the probability of being able to achieve these objectives is quite high, given the necessary financial support. Research of this kind, however, is also more directly involved in the production process than it used to be, and no country which wants to stay ahead economically can do without it.

Although the scientific community seems to be only remotely connected to the service sector, it is in fact an essential economic factor in shaping and supporting the information society. The work being done in research today will determine tomorrow's market. Almost all the industrial nations are aware of this fact, and there is a high degree of consensus concerning the key technologies so that the international competition for research money is becoming more ruthless. The American Strategic Defensive Initiative (SDI) is a vivid demonstration of the amount of money that can be involved in such long-term research projects. The basic research costs for the program alone are officially estimated at 26 billion dollars while insiders are figuring on three times that amount. Our example illustrates that even the USA cannot come up with that kind of money unless one can prove that such projects are of strategic value and be included in the defense budget. It also has the advantage of providing such research projects with a maximum of security and protects them against technological competition. As the race between rival research projects becomes more competitive, the risk of ending up throwing money away, in the event of failure or hold-ups, increases and the winners of the first lap will end up leading in the information technology sweepstakes by a wide margin.

This calls for new political measures. If politicians till now have been used to regarding research support and economic subsidies as two totally separate matters of strictly national interest, they now have to realize that the one without the other is worthless. In any event, as far as the research capacities of technological "middleweights" go, solo performances can turn into dangerous tightrope walks with the risk of falling headlong. If politicians until now have held to the principle of "money in return for results," in keeping with the good old tradition of public finance, they now have to spend billions according to the motto "money for hope."

Making research work a matter of international cooperation is therefore imperative for the development of the information society. At the same time, the importance of political control functions on a national level is on the decline. Both aspects create problems which, however, can be solved, if suitable measures are taken, and if the self-image of those in power is capable of freeing itself in time from the illusion of being some kind of global authority.

In view of this, one of the most important and urgent tasks facing European politicians today is to come to an agreement on a suitable and sensible approach to research policy. This question will be taken up in detail in the last chapter. Just recently the European Community has initiated a number of programs, such as the "European Strategic Program for Research and Development in the Area of Information Technology" (Esprit) which, despite their usefulness, tend to demonstrate the dilemma in which we find ourselves rather than promise a breakthrough. The 1.65 billion dollar budget for the whole program, namely, which is to be alloted over a period of five years for research in the area of micro-electronics, information processing, office and industrial automation, as well as software technology, is just a quarter of what IBM spends in one year (!) for research and development. The similar five-year "Research Action Program in the Field of Biotechnology," which the Council of EC Ministers established in 1984, has a total budget of 220 million dollars. The American chemical company DuPont, according to their own press releases, however, will be spending around 175 million dollars annually for genetic and biotechnological research over the next couple of years.

It shows that we are simply not dealing on a scale that can be described as a joint research effort. As long as agricultural interests continue to tie up the resources and obstruct the administrative channels of the EC and prevent an efficient research program from getting underway, the leading industrial nations of Europe will have to try to pool their resources and reach international agreements in as many areas of technologically orientated basic research as possible. Establishing research consortiums, multi-disciplinary research teams, integrated data banks, deciding on long-term goals of development in order to avoid overlapping and a waste of resources, as well as coordinating future projects with existing institutions and programs, are even more important than any individual aereonautics or space travel projects – although they have proved that Europe is capable of taking the lead in technology when it does decide to sit down and work together the European way.

The growing economic and political importance of a scientific and technological elite cannot be denied. It is a sure sign that we are in the process of changing from the macro-structure of an industrial culture based on empiricism to the micro-structure of an information culture dominated by theory. Does this mean that the writing is already on the wall? Are we already living in a society which can be characterized by Hans Lenk's question: "Will the new priests all be wearing lab coats?" Are the intellectual technocrats inexorably winning the upper hand in the power struggle, cooking up norm recipies in their bio-kitchens, busy replacing people with robots in their computer labs, and keeping a watch on the "il*bit*erate" average citizen in the central government offices? Will democratic politicians have to pay for their status as laymen/women and amateurs and give up their leadership and respective areas of competence to the scientific experts?

Such suspicions are just as short-sighted in my opinion as comparing computer performance to human intelligence or seeing information technology system networks as straitjackets for the intellectual potential of the individual. Specialized matters of scientific import are determining more and more the kind of questions being posed in our society today, but they are not providing the answers. Politics would be reduced to a reflection of technological and economic factors if those in power understood their job as simply organizing and regulating technological details, instead of creating a framework in which they could be made use of by society. Questions of orientation and meaning can be posed in abstract, scientific terms, but they can only be answered from an individual, ethical standpoint. Technological culture and value systems are not congruent, but rather counterweights of a system in which a tip in favour of the one or the other has to be continually balanced out by maintaining a dynamic equilibrium between the two.

We have already named a number of these counterweights: the striving for autonomy, upholding tradition and maintaining the substance of things, protecting the environment. Changes in social and educational policy can be added to the list as typical service factors. The way in which we divide the time at our disposal between work and recreation can affect our lifestyle and lead to changes in the demand for consumer goods. A lot can be done in helping to shape things if one understands and supports the role played by self-assertion in cultural and social life.

If the service factor is really to become a fruitful factor in shaping the information society, however, we will have to throw over board a number of cliches and correct misguided developments, such as our ideas concerning the following:

- that industrial laws and social regulations should ignore the differences between sectors, qualifications, and job features as much as possible in order to guarantee a " standardized and socially just" employment market;
- that the pay scale within the tertiary sector and between manufacturing and non-manufacturing industries must be brought into line with each other as well as
- the tendency to establish a service economy in the sense of a state welfare system providing protection from the cradle to the grave.

This kind of thinking is totally off base and cannot be justified by either an intrinsic logic or the interests of people. These cliches have become a standard element in many laws, however, and still have the support of influential lobbyists. This has lead to some very serious misdevelopments, the results of which we all complain about loudly, but whose causes we only halfheartedly try to combat. Among the most serious misdevelopments are the following:

- a black market economy which has grown to an estimated 75 billion dollars, the real winner in the scramble for profits in the last few years;
- an inflexible and bureaucratic employment market which can only see things in black and white terms: a person is either self-employed or on a payroll, you either have the choice of hiring people at exorbitantly high wages or resorting to illegal moonlighters, you either have a job which is practically impossible to get fired from or you are unemployed;
- a levelling out, concentration, and collectivization, particularly in labor-

intensive and people orientated services, which has lead to a massive drain on efficiency reserves and prevented us from taking maximum advantage of conceivable, worthwhile employment and social measures.

In the service industries it is especially apparent that we are facing not so much a "crisis of work" as a crisis of work regulation in the Federal Republic of Germany.

Everything is tailored to the model of a full-time job, which is the cornerstone on which not only individual incomes and social benefit claims depend, but it is also the decisive factor in determining what one has to pay in the way of taxes and other deductions. The most important laws regulating employment, social benefits, and income tax were conceived at a time when human manpower was at a minimum and a decisive production factor. Binding material rights and duties to a given occupation only makes sense for this kind of constellation. It is the only way of guaranteeing that anyone who wants to can achieve this type of job status and make it the basis of his existence. Shifts in the demographic structure of the population, along with the largely irreversible decisions concerning policy with regard to foreigners, however, have eliminated the scarcity of human manpower. The rapid strides in technological progress have also superseded people as the most important production factor in the manufacturing industries and replaced them with the need for additional capital and machines.

The logical response to this development would be to loosen the bond between employment contracts and social benefits, to place less emphasis on income and more on sales taxes, and to make working time more flexible. This would allow the individual to put him/herself in a position to be able to offer his/her labor at a competitive price but with a guaranteed income where it is most needed.

Instead, the social costs of wage earning have been increased, which has helped to precipitate the rate at which machines are replacing job holders and raised new barriers for hiring the unemployed. More stringent measures for the automatic withholding of taxes have been initiated which have narrowed the scope of action for making individual provision for one's assets and made the legal hiring of manpower even more of a luxury item. As a result, machine substitution, self-service, and moonlighting are becoming more worthwhile all the time. Legal options and collective agreements have reduced working hours wholesale, contributing to the strain on businesses and the social security system but not doing anything to make the employment market any more flexible.

In the past, the service sector has not been able to absorb the surplus of a depressed employment market, something which would have been especially helpful, both economically and socially, during the transitional period of the industrial sector. There was a loss of 2.7 million jobs in the agricultural and manufacturing industries between 1973 and 1983, as opposed to just one million new service jobs, sixty per cent of which were provided by the government.

Development took an entirely different turn in the USA and is often referred to today as the American "employment miracle," although it is no more miraculous than the German post-war economic miracle, which developed essentially as a result of the right market conditions, courage, and motivation. Fifteen million jobs were created in the United States between 1973 and 1983, almost all of them in the tertiary sector and only thirteen per cent by public institutions. Consultant firms, agencies, banks and insurance companies, private schools and health services, software manufacturers and distributors in transportation and communication networks provided new jobs for millions of people – and that in an economy which altogether was not growing at a much faster rate than the German one.

There is no use in trying to disqualify this success with handy catchphrases such as "McDonald's Economy." It is not only irrelevant, it also does not do away with the fact that over eleven per cent of the working population are in the meantime without jobs in Europe as compared to only seven per cent in North America. A static conception of the "job system" is the main obstacle standing in the way of people and preventing them from making the adjustment to the dynamic technological developments and new social needs. There are not enough bridges between work and recreation in Germany, not enough crossbreeds between employer and employee, and there is not enough faith in the creativity of the individual. On the other hand, we have more than our share of timidity, lack of imagination, and mania for regimentation.

There are signs, though, that things are beginning to change here. The need for stronger personal commitment in social fields, the renaissance of cultural activities, the popularity of clubs, the countless grassroots movements, the authentic or ostensibly alternative forms of living and working, the conscious-living groups, the thirteen million do-it-yourselfers, the computer games culture of many young people – these and other evidences of life seem to me to intuitively anticipate an individualized, decentralized, and deregulated society. The first intimations of change can be seen in the area of recreation where they can be most easily effected. Our non-working life is becoming a matter of increasing importance and value for us. It is not because Germans have decided that being lazy is a virtue to be aspired after, but rather that the narrow, limiting norms of the traditional forms and contents of work are increasingly felt to be inappropriate and restrictive in comparison to what one subjectively considers desirable and objectively important.

Since no one can permanently live in a state of conflict, however, which forces one to forget one's sense of values on entering the factory or office building, a new push for autonomy and partnership is slowly but surely making itself felt in the work world. Personal challenges are accepted and expected: More people are taking part in training programs and courses in adult education as well as offering suggestions and coming up with improvements themselves regarding their work. Organized worker participation, in the way it has been strictly, but also very formally, adhered to in dealing with assembly-line workers, is losing some of its dominating character while at the same time company agreements and group orientated self-determination are gaining in importance. More and more companies are offering flexible working times which allow for a great deal of individual variation within the framework of a monthly minimum of working hours. More than six million employees make use of flexi-time working schedules, two million people have part-time jobs and two hundred thousand work at temporary "not at a company level" jobs.

Longer sabbatical leaves, such as those now also possible in the civil service, are meeting more and more with a positive response. Only one in ten employees works till the legal retirement age of 65 – in 1973 it was still one in three. According to an inquiry by the National and State Social Benefit Institution, 95.8 per cent of non-civil servant employees of the government stop working before they reach the age of 65. A new social group of active senior citizens has emerged which is not at all content with just being socially passive consumers. Voluntary overseas service organizations and education centers have already discovered these "restless retirees," who are often highly qualified.

I have already pointed out that the trend towards self-employment continues to thrive. And it is not just software specialists, technologists, and restaurant owners who are taking a shot at being independent businessmen/women: Young people are establishing professional party services, offering their services as "errand runners" in the local papers, organizing successful early morning delivery service for fresh rolls, working as club house caretakers, tennis trainers, and holiday entertainers. I am convinced that they will branch out into a number of further service fields that are just as useful, communicative, and varied.

People are also giving more thought to new forms of cooperation in traditional service industries. In 1984 more than thirty contracting businesses joined forces in co-ops. By coordinating and dispatching the repair and remodeling work to be done by the different trades such co-ops are able to reduce the period of time in which home and condominium owners have a loss of rents. The contractors' commitment to work around the clock and on weekends, if necessary, makes up for the edge in competition that the moonlighters have while still conforming to the legal strictures of the market.

The trend is towards a relaxation of the strict segregation between one's work world and one's recreational life. This is certainly due in part to economic necessity, but it is also a proof of a pronounced endeavor towards autonomy. This could lead one day to a fundamental change and a fusion between the economic and social structures of our lives. At the side of the manufacturers, a new service middle-class could take its place, functioning as advisors and organizers and working with the aid of telecommunication. It could establish decentralized forms of self-organization in which customers would be offered individual as well as expert help in realizing their social needs and the desire to make things themselves. This would pull the rug out from under the black market economy.

The market for these kinds of services, orientated towards people and supported by technology, is a long way off from being saturated, as opposed to some other consumer goods. In fact, the demand for such services is just beginning to make itself felt. Our consumer market, entertainment industry, and cultural offerings are still one-sidedly youth orientated. In view of the rise in the average age of our population, however, this situation cannot continue forever. At present for every 100 people between the ages of 20 and 60, only 38 are over sixty years old; in just fifteen years 43 will find themselves in their "third lifephase."

It would be a mistake to think that these elderly people, who grew up with prosperity and who are used to technological progress and thinking in categories of self-realization, are going to let themselves be pushed into old people's homes or be satisfied with the crumbs of consumer entertainment. On the contrary, they will be a central factor in cultural and social decision making. Their value preferences (health care, experience with nature, educational background, keeping up social contacts, family ties) will change the structure of our consumer markets and services. To ignore the wealth of experience and time at their disposal by sticking to an inflexible policy of forcing elderly people to retire as early as possible and locking them away in old people's ghettos will have to be recognized as a grotesque waste of very valuable resources. To hold to this policy of waste is not only impossible financially, in the long run it is also a gross miscarriage of justice for a society that likes to consider itself humane.

The sooner we start differentiating between individual needs the better off those concerned – and that means all of us eventually – will be. Our social welfare system is in need of overhauling, especially when it comes to how we deal with older people. Putting someone in a home and in-patient hospital care should only be taken as a last resort, only when they are absolutely necessary steps, both from a humane and medical point of view. Government policy concerning the subsidy and reimbursement costs for institutions offering these kinds of services should be brought into line with this new outlook. Neither eventual inconveniences for families nor the capacity plans of those responsible for social welfare should be the decisive factors in determining the direction our social policy should take. We should orientate ourselves alone on the ideal of a fully integrated generation which has the right to be needed and taken seriously, even after the age of sixty.

A graded system of welfare, based on consistent principles of decentralized responsibilities, is therefore just as important as building up a voluntary structure of care service which is consistent with the abilities of the senior citizens themselves. Out-patient care and halfway houses could give older people the opportunity to actively support and help other elderly persons. Meals-on-wheels services could be organized by charitable organizations with the help of volunteers, housewives, and senior citizens. Community work in the field of landscape care and environmental protection could be a chance for youth organizations and qualified senior citizens to bridge the generation gap and experience a sense of solidarity with one another. Former doctors, physiotherapists, and retired physical education teachers could run health training programs in community centers. Trips, recreational activities, and educational courses geared to senior citizens more or less require the participation of older people in the "doer" role, and not just as passive consumers. Maintenance and repair work for the low income elderly and in senior citizen community centers could be taken over by former skilled workmen/women without being stigmatized as moonlighting.

On the other hand, those who rush to the defense of the existing laws regulating trade and industry with the sword-in-hand in a crusade of job "protectionism" overlook three important aspects of the problem:

- Some of the examples mentioned above and other forms of individual and group self-help are already being practiced, whether as public model projects or in the legal no man's land of a degrading and forced secrecy.
- Constructing artificial barriers for taking advantage of lower-priced services, despite demand and social legitimation, does not increase the demand for high-priced commercial services but rather simply speeds up the process of technological substitution (viewdata service, telemechanics, telecommunication, and remote control) and makes authentic moonlighting work more attractive.
- Last but not least: What makes the new forms of service, in which the older generation could be integrated as active participants instead of as simply passive consumers, especially attractive is their potential for an in-

teractive division of labor with the professional economy. Experience has shown that this could very well give new impetus to demands that otherwise would not arise.

A lot of work still has to be done before these and similar examples for gradually making the switch from an active worklife to retirement – basically a very natural way of proceeding – can be put into practice. Not only do a number of legal aspects have to be cleared up, but it is also important to avoid a direct clash with those representing group interests. The conflict will not be mitigated until we risk taking the first step and do away with 65 as the mandatory retirement age. This can only happen within the framework of a general reform of the laws regulating, not only working hours, but also our system of taxation and social insurance. The pressure for providing a guarantee for the payment of social security benefits beyond the year 2000, though, and a visible growth in the political power of senior citizens will force things to be brought to a head by the nineties at the latest. In the meantime, it would be advantageous to accumulate as much practical experience as possible in suitable economic and social fields of activity with "part-time pensioners." In any event, the provisional character of the early retirement regulation, valid till 1988, should be kept in mind by the population (and not be blurred by extensions, as have been proposed for the civil service) as simply a means of avoiding the temporary unemployment problems we are now facing as peak numbers of baby-boomers crowd into the job market.

Personal services, and not just in the social field, are on the rise. In this respect, Bell's and Fourastiés' thesis that scientists and engineers will also play a leading role in the information society in terms of quantity, is probably not correct. On the contrary: Aside from technology, production, and consumer buying, the demand for services is expanding rapidly. This trend cannot be categorized in terms of material goods but rather has an intrinsic value in the sense of personal concern, interpersonal contact, and caring for the environment and culture. Whether this demand, which has not been satiated by any means, will also have a positive influence on the job market depends on how much freedom the new market will be given.

Many of these activities fall into the realm of recreation – club membership activities, cultural entertainment, sports, adult education. They are highly suitable for forming limited contracts if the alternatives are either long-term unemployment or being forced to wait indefinitely for a non-tenured position. They could partially replace family income if, for example, one wage-earner should fall flat and thus prevent social welfare benefits from automatically having to come to the rescue. A prerequisite in all these cases is that such work can be carried out flexibly, in terms of time, and in the form of a kind of "recreational self-employment," which is not subject to strict and complicated taxation and insurance regulations.

Being able to offer one's services legally as a part-time self-employed businessman/woman, while being regularly employed elsewhere, should also be given more consideration. In this way people could be given a chance to check out the possibility of establishing themselves in their own small businesses.

Furthermore, new forms of commercial self-organization and suitable adjustments in terms of job content deserve support. Artisan collectives, similar to the backyard shops that have already been established in some large cities, are a good way of cutting down costs and increasing efficiency. Most communities own warehouse and office space which is not being used, and which they could rent to young business people at low cost. These business collectives could find their best customers in hobby do-it-yourselfers, by not only selling them material, but also by offering them qualified advice for a number of overlapping areas of skill. The skilled knowledge of the craftsman/woman and the enthusiasm of the do-it-yourselfer – together they could help each other, the one in earning a living and the other in saving money. Since the going rate for qualified labor by the time you add up social benefit payments, overhead expenses, and additional costs, is beyond most people's budgets, it is important to differentiate between the indispensable skilled knowledge of the craftsman/woman and the actual labor which often requires a great deal of time but which can be carried out, at least in part, by non-experts. The skilled trades, if they continue developing in this direction, could turn out to be one of the prime examples of using one's talents in the new people and information orientated field of service.

A fundamental change of structure will also have to take place in the agricultural field. Although the number of people employed in agriculture has been on the decline for the past decades, the farms still in business – a result of the economic absurdities of the EC's agricultural policy – have been producing increasing surpluses and causing economic and ecological damage which is becoming all too obvious. The competition is threatening to entirely kill off smaller, out-of-the-way farms or those run as a means of earning additional income. They should be given a new perspective in work involving the care and maintenance of a balanced landscape. Independent, although government supported, work as private landscape caretakers would help to reduce production capacities, the environment would benefit directly, and pointless (in the sense of producing surplus goods) subsidies would be done away with. The "environment-entrepreneurs" would perform a public service which the government till now has had to more or less carry out on its own, but which has been expensive and done only halfheartedly because of the conflict between conserving nature and maintaining a livelihood for farmers.

The boundary between changes that are possible or even desirable in the traditional production and service fields and the so-called alternative workshops and businesses is wide-open, as one sees. This is by no means a matter of chance. Both developments place a strong emphasis on personal commitment and have an unorthodox attitude towards disposing of one's working time and the wish for determining one's own job content in common. Cultural preferences influence economic structures – the heterogeneity of the tertiary and quaternary sectors is the strongest proof of this.

There is a great deal of confusion, however, in trying to judge between what is (still) "economic" and what is (already) "anti-economic." The conflict is even reflected in the matter of terminology: The term dual economy is used both by those who see a hierarchy free eco-production as the only alternative to an industrial lifestyle, as well as by those who want to indicate the exact opposite, namely the fear of the information society splitting people up into small, powerful elite groups and a large section of menial subordinates.

In my opinion, there is a tendency to polarize and erect artificial barriers between the two standpoints which could very well be meshed together. A system of values which strives for non-alienated work, an environmentally friendly form of production, direct solidarity, autonomous time disposition, and moderate consumer buying, as we have seen, can easily be interpreted as the worthwhile aims of an information society in which the creative potential of modern technology is made the most of. This cannot be done by closing our eyes to micro-electronics, but rather by rejecting assembly-line work as the final state in the division of labor between man and machine.

The belief, however, that "back to nature" means going "back to preindustrial forms of production" is deeply rooted in most alternative enterprises. This leads, as a series of articles in *Zeit* magazine vividly illustrated, to a grotesque and completely unnecessary self-exploitation. The original vision of establishing a lifestyle in which recreation and work are an integrated whole gets lost in a harassing grind of work with inadequate devices barely covering minimum living costs: "The pronounced transfiguration of work done by hand in the anti-economy has resulted in drop-outs, who otherwise in a regular job could not hammer a nail in the wall without banging their thumb, are trying their hand at being carpenters, metalworkers, and plumbers." As long as this does not change and an amateur attitude towards technology and finances is regarded as a proof of an alternative lifestyle, the "anti-economy" will continue to be simply a satellite, dependent on the functioning of the official economy. This is the case, despite around twelve thousand alternative projects, among them more than a thousand bio-farms, and despite some very worthy ideals.

Just as the ecology scene has blocked the way for realizing its full potential by placing a taboo on taking advantage of the opportunities offered by technology, the high-tech economy, on the reverse side of the coin, is pursuing a non-material taboo policy which has led to a de-valuation of the post-industrial technology in the true sense of the word. Traditional patterns of work are still retained although flexibility, both in terms of time and space, would be possible. The management structure is left untouched, although vertical and horizontal room for autonomy could be created. Concern for environmental issues often still does not surface except as a reaction to public pressure.

The government itself does not deal with the matter much differently: The regulations regarding working time rules, trade, social benefits, and taxation have not been altered to keep pace with the changes and make the most of them. Recreation, the striving for autonomy, social or cultural commitment, decentralization, and the infrastructure of the information technology could function as a buffer and create a counterbalance between the supply of and the demand for work.

Work and recreation, gainful employment and retirement, self-employment and dependency all continue to be entirely separate spheres of activity. As a result, where variability in terms of time is called for, flexible machines are more suitable than people who are prevented from shaping their own individual work and lifestyles as free economic citizens by a casuistic system of regulations. The priority granted the automatic withholding of taxes makes legal employment disproportionately costly and moonlighting very tempting, consumer buying, especially in comparison with personal services, a very good deal. To state the matter blatantly, the skilled laborer has to pay such high taxes so that the moonlighter can buy his materials cheap and the Japanese companies can sell their video recorders at a good price. In its current form our tax system, without a doubt, favors the polarization between a concentration of business and capital, on the one hand, and a black market economy, on the other; it is the middle-class who usually ends up footing the bill in the form of astronomical labor costs and high taxes.

On top of the unbalanced system of taxation, there are the exorbitant social contributions so that an employee ends up costing his/her employer twice as much as what he earns in wages before taxes. There would be nothing to object to in this if these contributions really served to support the really needy, low-income or large families. They also benefit, though, people who have long since been making enough money that they no longer need collective support on such a large scale. In such cases setting a limit on the amount of income liable to taxation and social contributions would not only be warranted in terms of social justice, but it would also give the individual more freedom of choice in deciding how to make best use of his available income – whether it means taking out additional private insurance policies, investing money for the future, or raising his/her present standard of living.

We are now facing the following dilemma: Although the alternative economy is striving to establish a (at least in part) progressive system of values, their backward attitude towards technology prevents them from achieving their goal. The "official" economy, on the other hand, makes use of progressive technology but is handicaped in developing a corresponding progress for society as a whole by a system of values which is constrained by the dictates of a static industrial society and which is reflected to a large extent in the framework of our economic and social legislation. In both fields information technology has caused, unnecessarily, a syndrom of fear and repression which, although in vain, people have reacted to by attempting to ignore the situation or dropping out altogether.

Are the two positions really irreconcilable? Let us recall the following to mind: We are already seeing far-reaching changes being made today in our system of values which make the inadequacies of our current employment system painfully apparent and reveal the potential in making use of the information technology in order to achieve more self-determination in the work world. As we have seen, the information technology plays a very important role in overcoming the old antagonism between industry and environment. Irreplaceable personal services are becoming more and more important in alleviating social evils. An overtaxed and inflexible employment market and the associated collective social security system produce an enormous market potential for either full or part-time self-employed businessmen/women in (almost) any kind of advisory, go-between, or supporting capacity. Everything, in my opinion, points towards having the technological evolution go hand in hand with an innovation of society as a whole by liberalizing anachronistic regulations which have been stripped of their original protective function. This would not only allow for more freedom and self-realization, but it also calls for more personal responsibility for society and the environment. Learning that the one determines the other and that communication technology is only worthwhile if communication in its entire complexity is esteemed, could be one of the most important and stabilizing elements in shaping the information society.

In answer to the criticism, that I am painting a far too rosy picture of future strategy – a humanization of automation by means of autonomy – I would like to emphasize that there are, of course, a number of elements involved in social change which cannot be calculated in advance. We are not in a position today, for example, to be able to judge what the going rate for simple services will be if price negotiations are worked out on an individual basis rather than in strict keeping with collective bargaining agreements. At first a kind of duality between well-paid technologically orientated and relatively low-paid person orientated services could very well occur.

This would not be very different, though, from the situation in which we are now. Model calculations indicate that the incomes of the lower-wage earners – in the service sector these are for the most part people working at unskilled and errand jobs – barely amount to much more than they would be entitled to if they had to live on welfare. This is especially true for large families where there is only one wage-earner. If one does not want to mince matters, it clearly reflects the poor opinion such work is held in. A civil servant assistant with two children earns 1,200 dollars after taxes from which he has to pay the rent and his health insurance. An unemployed welfare recipient with the same number of children gets almost 1,000 dollars a month, including supplementary payments and rent reimbursement. The "surplus value" placed by society on work versus non-employment in such a case hardly amounts to a hill of beans.

The job and social duality which many people are afraid of in connection with a society shaped by information technology already exists in our industrial society without being especially apparent. There are reasons for this: First of all, the consumer ends up paying so much for any kind of service, due to the astronomical social contribution payments involved in regular work, that people tend to feel that such work is overrated rather than underrated, and rightly so from their point of view. Secondly, as we have seen, the difference between working and not working at the lower end of the income scale is so minimal that people are not motivated, from a material standpoint, to work at such lower paying jobs in order to make a living. In other words, the extent to which having a job or not makes a difference in one's income is obscured by the welfare state's twofold strategy for redistributing the wealth.

The recommendation of stimulating the private service industry by relaxing the regulations governing work, social welfare benefits, and trade might lead to greater disparities between wages and income in the early stages but, aside from making the present state of affairs simply more transparent, basically nothing would be changed. The possibility of part-time self-employment, parallel and second jobs, differentiated wage structures, and (of course taxable) work in one's free time, though, would give people an opportunity to legally make money. This could lead in a very short time to individual "total incomes" which would be much higher than the present level of wages.

A tax policy that was more consumer and less wage orientated would give human labor, especially in its technological and indispensable components, a real "concrete" market value again which would distinguish it from non-work and moonlighting and make it worthwhile. A greater degree of self-determination would increase motivation, as numerous studies dealing with value changes have shown. This would certainly not only be economically beneficial but also have a positive influence on the social and cultural sector. A policy which gave more heed to freedom in the employment market, therefore, would not only be economically feasible but also a worthwhile and progressive form of social policy.

In our discussion of the conflict between services and the new technology so far we have been concentrating on the two counterpoles in a broad range of activities – the scientific and technological protagonists on the one end and the pronounced personal services on the other end. Since the interdependency in these fields between technology and culture – in the sense of the political challenge as well as the strategies of the individual and society for coping with the change – is especially apparent, this type of comparison was the obvious approach to take. One should keep in mind, however, that we are of course only dealing with segments of an almost incalculable range of various jobs and services in the tertiary and quaternary sector. What the total impact of the information and communication technology on trade, transportation, the credit business, insurances, the supply of energy, tourism, education, health, and the various production orientated services will be, can neither be expounded within the framework of this book nor, as I see it, is enough adequate information available at the present time for making such an analysis at all.

Two general tendencies appear to me, however, as characteristic for the logical development of the new technology: First of all, a greater demand for technology, system, and production orientated know-how, and secondly, the need for more intensive personal counselling generally in the service field. The manufacturing and service sectors are being brought more and more into line with one another. For the first time in 1983 information technology, as a branch of the mechanical engineering industry, had the highest production output, overtaking the traditional machine tool branch which till then had been the industry leader. More service orientated user know-how is thus becoming the prerequisite for a successful production and sales strategy. In the machine tool branch itself the importance of electronical services in the form of sophisticated, flexible programming is growing in proportion to the complexity of the production systems. The nonmaterial "raw material information" is taking over the traditionally mechanical production, reshaping it, and increasing the percentage of preliminary and auxiliary services involved in such work.

A similar trend can also, of course, be seen in the opposite direction: Modern business services require a lot more information concerning technology and machines than they did in the past. This applies most obviously to their organizational structure; it is also becoming a more important factor in determining what kind of work is done. Software companies and (to a certain extent professional) information data banks are being faced with the growing demand for individual solutions to problems which they can barely cope with due to the lack of trained personnel. The still prevailing overwhelming predominance of foreign, in particular, American data systems and standard programs on the German market is not something we can afford to ignore. It fosters economic dependency and cultural conformity which hamper independent development and, above all, often prevent other new technologies from becoming widely accepted. Management consultant companies will also have to have more information technology know-how at their disposal if they want to stay competitive.

In a highly integrated national economy things inevitably develop according to their own laws. The banks, for example, are being faced more and more with the problem of having to judge the market chances of their customers' technological innovations in advancing business credit. Due to a lack of qualified personnel, they are rarely in a position to make such judgements, at least this has been the case until now, so that it is relatively easy to get credit for investing in new industrial goods but not for developing or purchasing new innovative products. Efficiency measures make sense in terms of dollars and cents but not a new measuring device. The average bank credit officer and the young electronic engineer who wants to set up his own business speak different languages. This has had a disastrous effect on development, particularly for middle-sized businesses. To a certain extent the government and business community supported advice centers for technology transfer serve as valuable "translators;" more product know-how (which very feasibly could concentrate on certain target groups), however, would be a highly desirable form in which financial institutions could serve the economy as a whole.

The service society will have a difficult time developing into an information society if it fails to get a new, comprehensive grasp on the role played by information and consulting services. Part of the conflict in our industrial society can be traced, without a doubt, to the discrepancy between specialized knowledge and inadequate means of communication. Our society has an enormous fund of knowledge at its disposal, but it is buried in the labyrinth of out-dated training methods and organizational structures. A society like the Japanese, in which people have been brought up for centuries to value group consensus and an "inner openness," obviously has a much easier time in adjusting to the demands for exchange and agreement of an information society. This is one of the reasons for their success. The willingness to compromise, patience, the ability to learn, and personal commitment are indispensable criteria for using the new technologies in both an economically and socially worthwhile manner. This indicates, though, that a society which cultivates conflict and delimitation for its own sake can never become an information society. It will always be moving in that direction but will never reach its objective because it does not comprehend the cultural and social message of the new technology.

West Germany should not be among the nations that get lost on the wayside. We still have time to reflect. Time is running out though: There are around three thousand working days left till the end of the century, and every single one of them counts.

Old Values – A New Order

Chapter 12
New Vitality for Old Values?

Carrying Enlightenment to Completion

In the course of our reflections till now we have often touched upon questions of value and orientation; individual and social "snapshots" within the context of structural change. We have also frequently talked about the political consequences. At this point we would like to attempt to reconcile the different aspects of the change in values, the economic and social structures, and the technological development within a general framework of a public policy.

I would like to emphasize the word *attempt* in this context. The problems confronting us are enormously complex and the politicians, as non-experts, are faced with certain limitations in meeting the challenge. Wanting to achieve just a general outline of developments requires not shying away from playing the dilettante in many fields which the experts themselves often do not have fully in their grasp.

Nonetheless, I would rather be critisized for concerning myself about subjects in which I am not a specialist rather than later being accused of having failed to do what lay in my power and of having ignored the challenge in fear of the complexity of the social problems involved.

The former offers at least the opportunity of not only making mistakes but also of acting as a starting-point for initiating and effectively developing worthwhile political measures. On the other hand, by consciously closing our eyes to the developments around us, we run the risk of sinking into a slough of self-satisfaction – just about the worst thing that can happen to a politician's self-image.

My thesis, which I will substantiate below, reads as follows: Values which emphasize the importance of man's social existence will play a much more important role in the information society than they did in the industrial society. Old-fashioned values and virtues could unexpectedly undergo a revival in the course of this change. In order for this to happen, though, a number of patterns of social order will have to be freed of the political and dialectical polarity in which they currently find themselves confined. In terms of a new social policy they will have to become more personalized and from an ethical standpoint made a matter of global concern by establishing a sense of comprehensive interrelationships. This cannot be successfully accomplished in confrontation, but only in cooperation with the new technologies.

A few preliminary remarks regarding the shift in values which the industrial society produced are called for, however, in order to get a better grasp of the position in which we now find ourselves, comprising as it does the starting point of reference from which a development of this kind must be carried out. In my opinion, there is no mistaking the fact that the individual's potential for development has been curtailed by the standardized systems of industrial life established in the 19th and 20th centuries. Among these standardized systems are the highly inflexible division of labor between man and machine as well as the rigid regulations concerning working time and the collectivized social welfare system which pay scant regard to life cycles or family life. Under the pressure of the social structure and the job system the opportunities for self-realization have become consigned more and more to the private sphere of recreation and consumption.

Seen in the larger historical and philosophical context, however, this disjunction appears to be more of a paradox rather than a logical result of developments. It was in fact the trade rights of the bourgeoisie, in the form of town and guild charters, which laid the cornerstone for the development of civil rights and liberties. The realization that the talents of skilled craftsmen and merchants, at least in times of peace, were more important than cultivating aristocratic traditions, inexorably led to the erosion of the feudal system. The German bourgeois won his liberty, not by means of a revolution, but by the fruits of his labor.

Also in view of Germany's contribution to one of the most important periods in Western philosophy, the Enlightenment, there are – if I see things right – enormous discrepancies between philosophical ideals on the one hand and the realities of the industrial society on the other.

"Daring to know," the prerogative of reason, was not meant to be taken as an end in itself, but rather as a means of freeing oneself of "a self-imposed immaturity" (Kant), and thus paving the way for the responsible use of individual freedom within the framework of a just public order. The rapid development of the natural sciences and technology, the "applied enlightenment," so to speak, produced social structures, however, which did not grant the majority of the population either more freedom or more justice. The progress of civilization, as a result of enlightened reason, first found expression in the form of machines and changes in production processes and not in a transformation of social ethics. In time, of course, industrial progress did lead to more freedom and human dignity for mankind. The course of development extended over the timespan of several generations, however, and a number of factors were involved in the complex process: The pressure of interest groups (in other words by the subjection of the discernably weaker power of the individual to the more powerful collective amalgamation); the prudent policy of those in power (Bismarck!), which sought to nip the potential power of revolutionary forces in the bud; and the gradual rise in the economic standard of living as a result of increased productivity.

We are still being influenced today by the experiences made in this long and thorny process. Technology and management are still regarded as enemies rather than as partners in the human struggle towards autonomy. The idea of an employee finding his only protection against exploitation in the power of collectivism sounds like the sentimental effusion of a bygone class struggle. Nonetheless, it is still a standard feature in the overall body of industrial law and thus has a direct influence on the norms regulating our economic system. Although other forms of regulating weekly working times would probably give people more time for recreation and family life and be better suited from an organic standpoint for the purpose of relaxation and recuperation from the stress of work, the workfree saturday is clung to with tenacity. The only plausible explanation for this phenomenon is no doubt a traumatic carryover from the early days of capitalism in which a minimum of opportunities for regeneration had to be bitterly fought for.

In other words, industrial change did not take place within the framework of a homogeneous, but rather an ambivalent system of values. The triumph of reason went hand in hand with a substantial loss in autonomy, which not only pulled the rug out from under man's social position but which was also very disorientating for his sense of "ethical self-conviction." Reason was reduced to a matter of intellect so that the moral precept of reason, formulated by Kant as a categorical imperative: "Act that the maxims of thy will could always function as a principle establishing universal law," did not find any echo in the practical business of life. It was overlaid by a fascination with a liberalistically orientated rationalism. When this attitude led to socially unacceptable conditions the approach taken towards making amends was not one of a different, namely socially ordered freedom, but rather one of a collectivization of an individual sense of ethics. The incomplete enlightenment, incomplete for the reason that its potential, but not its actual realization, was of central importance, only ended up considerably deepening the rift between the individual and society.

Nevertheless there is no doubt that historically legitimate reasons existed for establishing large standardized social systems. After the "shell as hard as steel" (Max Weber) was erected around industrial capitalism, there was no other realistic alternative. It would have been futile a hundred years ago to try and establish self-determination on the job and social welfare on the general principle of the freedom of the individual. The pressure of events and poverty had set an economic process in motion, with a momentum of its own and not adequately foreseen. The question is whether these same conditions still apply to our situation today. In my opinion the injunction of the enlightenment, to free man of the restrictions limiting him in the development of his full potential, is still waiting to be realized.

The citizens of our industrial society are very aware of the large discrepancy between their professional and social dependency and the value attached to the freedom of the individual. They are in search of a means for compensating for this discrepancy and generally find it in their recreational life. Recreation has become an end in itself for many people, their only chance to make up for the lack of self-realization in the other areas of their life. This partition of freedom and self-determination, which has its roots in industrial history, has led naturally to a number of paradoxes which can be seen wherever one looks: The new technologies, which are perceived as threatening in the area of work, are very popular as consumer articles. While conforming to the virtues of self-discipline and compliance at work, the emphasis after closing time is on respecting the right to do what one chooses. The thrill of personal risk, no longer a factor involved in making provision for the business of daily life, has its outlet in the form of flamboyant displays of personal courage in the area of sports and other challenging recreational activities. Taking on personal responsibility, which people are reluctant to accept on the job and only under the provision of certain guarantees, is considered a "matter of honor" in community and club organization work. The thoroughly standardized work society, as Bernd Guggenberger once described it in the *Frankfurter Allgemeine*, produces a kind of Dr. Jekyl, Mr. Hyde effect, a change of personality after leaving the office or shop.

Many of our legal and social structures, which were conceived during the last century with very different technological conditions in mind, need to be critically re-examined in my opinion. This is the only way of being able to judge just what social and political measures have to be taken in order to make the best use of the information society, i. e. to increase the quality of our lives. We cannot hope for a change for the better by limiting the information society to simply restructuring economic and administrative systems in keeping with the new means of information technology. This kind of one-sided thinking would inevitably lead to new social conflicts and to a lack of cultural and ethical vitality. After our experience with the industrialization phase, we should be doubly aware of the real danger of again trying to patch things together with a formal set of norms and standards when an individual and creative approach to the problem is called for.

The aim we are striving for is necessarily an ambitious one. The information society will only be regarded as a positive step in the development of human progress if it succeeds in controlling the complex economic, ecological and social processes involved with more consistency than the industrial society did. It must be in a position to help people overcome outdated restrictions, not at the cost of their fellowmen, but by the effective use of machines.

In a thriving information society, for example, people would no longer be directly affected by the monotony of mass production. With computer controlled robots directly responsible for supplying the material and for manufacturing, assembling and loading the products, a factory would become nothing more than a service enterprise which included an automated production division. The human field of activity would not be characterized by direct contact with the material product – which has not been the case since the division of labor along the lines of Taylor – but by the ability to flexibly control processes, whether at the cross sections between the flow of information and material, in the surveillance of production systems or in the development, administration, sales, or marketing divisions. This calls for a high degree of decision-making ability and a need for knowing how systems function on the part of those working well below the traditional management level. This in turn has less to do with formal job qualifications as with personal flexibility and ability: this is the determining factor in the synthesis of technological and human orientated services. This brings us face to face with one of the most puzzling inconsistencies of the industrial society: despite complaints, most people seem to find it easier to operate a machine rather than be of service to their fellowmen. When human performance is no longer forced to keep in step with the tempo of machines, however, the personality of the individual will automatically regain its central importance as a distinguishing mark of qualification.

In an empirical study published in 1984 by the industrial sociologists Kern and Schuhmann ("Das Ende der Arbeitsteilung? Rationalisierung in der industriellen Produktion" [The End of the Division of Labor? Rationalization Measures in Industrial Production]), the authors pointed out that our ideas concerning management and production are undergoing an important change, especially in the central fields of industry such as the automobile, mechanical engineering, and chemical industries. Over and above automation and rationalization measures, the idea of an "integral employment of work capacities" and making better use of employee qualifications are becoming increasingly important. Organizing work along the lines of Taylor with its strict division between planning and actual operation no longer makes sense, especially at jobs that involve a large amount of capital investment, because it results in behaviour which does not do justice to the sophisticated technological facilities whose operation requires a high degree of personal responsibility. In other words, there are sound business management reasons for assessing the qualitative value of human labor in proportion to the power of the mechanical production apparatus. This line of thought has given rise to new forms of operational organization, such as the "quality circle," which organizes the work force into relatively autonomous, interdisciplinary groups.

It seems as if individuality is gaining ground in the work process, particularly in the pre-production and post-production phases of operation. Human resources management is playing an increasingly significant role and, together with a flexible attitude towards working times, could gradually bring the dissonant value systems of work and private life more into line with each other. The most obvious paradoxes would be eliminated in that computers and robots would take over the work processes which are least adapted to human nature. Man could regain more in the way of inner freedom and balance. Then, however, it would have to be definitely shown what man can achieve after finally having found an "escape form his intellectual confinement."

Chapter 13

Decreasing Demands on the Environment Means Making Sacrifices

What Can We Do? What Should We Do?

The most important contribution of the new technologies towards the history of civilization is probably the opportunity they offer for developing unused human potential. At the same time it is the greatest stumbling block which has to be overcome in introducing the new information technology. Shirking responsibility in anonymous organizations rather than taking things upon oneself has become the determining social pattern of our times. As full of contradictions as the industrial society is, one can live quite comfortably in it as long as one can come up with a few halfway plausible arguments for shifting the responsibility for solving the problems onto someone else – whether it is the government, the environment, foreign countries, or interest groups.

For a long time, for example, there was no other alternative for building up industrial structures than exploiting the raw material reserves of the Third World. The government was then responsible for providing a certain degree of compensation in the form of trade agreements, foreign aid, and cooperation in international organizations. As we know, this was only achieved to a very limited degree – a few countries were able to pull themselves up by their own bootstraps and catch up, but in overall terms the gap between the rich and poor nations continually widened.

Nowadays basically every highly industrialized nation has enough scientific and technological know-how at its disposal that it could switch over to a national economy based on advanced technologies, demanding a minimum of raw material, and production methods requiring intensive research and service. From a practical standpoint, they would easily be in a position to let the developing nations take over a large percentage of the heavy industry and the low-technology products while at the same time providing them with the necessary environmental protection technologies. That this is not the case is not due to technological reasons but is rather a symptom of fear. We are afraid of introducing the new technologies with the necessary directness and consistency and shy away from demanding a personal commitment to the structural change from our citizens – in other words potential voters. Non-competitive industrial branches should be dissolved by steadily cutting back on subventions, and research and development efforts should be drastically increased. Further training should be granted the status of a "social duty" and be established as an integral element in the collective bargaining agreements between both sides of the industry. The need for developing a scientific elite should be reflected in the restructuring of the budget and educational planning.

Small steps have already been taken in this direction, but the efforts fall far short of making any waves in the placid waters of the political status quo. From the standpoint of domestic policy this attitude is understandable but unsatisfactory when measured against the ethical criteria of what has to be done. In failing to realize the potential of their economic structures the highly industrialized nations are blocking the chances for development in the Third World. Delays in adapting to new structures always entail the need for more stringent protectionist measures which automatically hit the weakest members in the international competition for economic survival hardest. The responsibility for latent domestic conflicts is thus shifted to others without there, however, being any objective need. The fact that conserving outdated structures with the aid of government subsidies is in direct contradiction to a policy of sound economics does not help matters. In the end, as we know, the countries will have to pay for their economic immobility; on the one hand with a decrease in the rate of productivity, and on the other hand with a drop in the demand for their products by the very same foreign markets whose progress towards industrialization was delayed by means of such protectionist measures. Those ending up footing the bill, in the form of unemployment, though, will be the very citizens who were supposed to be saved the trouble of making any kind of personal adaptation.

The nations that are capable of using and developing the new technologies are under obligation, not only according to the laws of reason but also from a moral standpoint, of using the opportunities available to them. Although questions concerning the possible consequences of the new technologies are important, they should not be used as an alibi for avoiding technological and economic adjustment problems at the cost of the social development opportunities of others. Whether man should be allowed to do everything he is technically capable of is a question which is frequently and passionately discussed in our culture. Whether, though, a share of the blame for the misery of others falls on us in failing to take full advantage of the opportunities available, whether being able to do something places us under moral obligation to actually go ahead and do it – these, admittedly uncomfortable questions are generally ignored.

The progress being made in technology is making it more and more difficult for us to continue to ignore existing conflicts with a clear conscience. Even those who do not feel particularly under moral obligation to help the desperately needy of the Third World will not be able to exonerate themselves on the grounds of ignorance for much longer. Not only the future of our environment but also that of our children and grandchildren is inextricably connected to the willingness of the current generation not to take the easy way out. "Stackless" production methods and miniature products are environment protection at its source. The more intellectual activity – knowledge, learning, communication, creativity – acquires a value of its own the less need there is to make use of other human resources. Information as a production factor functions as a substitute for natural raw material and tends to reduce the importance of products with little or no information content in the overall economic scheme of things. A ton of highly integrated circuits represents a value of approximately 100 million dollars. To produce the same amount of value in the steel industry more than 200,000 tons of crude steel would have to be formed into metal sheets half a centimeter thick. According to the emission statistics of the Second Immission Protection Report of the German federal government, 0.5 kilograms of dust, 0.13 kilograms of sulphur dioxide, 0.6 kilograms of nitrogen oxide, and 13 kilograms of carbon monoxide are produced in manufacturing one ton of crude steel. Multiplied by the factor of 200,000, this is the price in terms of the environment which we have to pay for outdated manufacturing processes in comparison to the new technologies. Furthermore, a ton of concentrated information technology encompasses an enormous potential of application uses which in turn save resources – for example being able to store the knowledge of whole libraries without having to cut down a single tree for making paper. (This does not imply that we will no longer have to know how to read, on the contrary, more and not less will be read.)

Naturally manufacturing microchips uses energy that has to be generated first and chemicals are needed which require strict security measures. Several Silicon Valley incidents have become known which indicate that the necessary precautionary measures have occasionally been disregarded; under these circumstances, as in any improper use of technology, environmental or even health hazards are inevitable. If used properly, however, and this is the crucial point, the amount of harmful substances emitted in the process of manufacturing microchips is well below the levels which prevail in the conventional steel and heavy industries employing the most modern filter systems. The security standards of Germany's industry are renowned worldwide; the chemical industry in particuliar registers fewer industrial accidents than any other branch of industry. The promise the new technologies therefore hold out of being environmental protection put into practice is well-founded and verifiable, at least for the Federal Republic of Germany.

The gradual replacement of natural resources with our own resources of intelligence can only be accomplished, however, if we do not fail to meet the intellectual and ethical challenges involved in this changing process. High-tech procedures also demand intellectual high-performance; if we want to relieve the strain on nature we will have to place more demands on ourselves – there is no way of getting around this.

If the concept of a contract between generations, which has always been seen in connection with social politics, is ever to attain authentic legitimacy, then fundamentally in this field. Whether future generations, in view of their disadvantageous position in terms of age structure, will fulfill their social obligations to those currently employed as set forth by the prevailing system of social security, is quite open to question. On the other hand, we can be certain that our descendents will want to roam through the forests, swim in the lakes and not be forced to emigrate due to economic want. Leaving behind an environment worth having and creating an infrastructure capable of development therefore constitute original, binding obligations between the generations.

It would be a fatal mistake to believe that things do not depend on us as individuals because the government will and has to make it its business to fulfill these obligations. Things do depend on each and everyone of us. Our constitutional rights of freedom and liberty do not allow the government to force anyone to acquire the necessary qualifications in order to be able to participate in the creation of non-polluting and internationally competitive products. It cannot prevent a company from transferring its base of operations to foreign countries if suitable workers cannot be found at home. The government can have underground glass fibre lines laid out but whether they end up being an investment for the future or simply dead capital does not depend on the officials in charge but on the man/woman in the street.

Social security – which we will speak about in more detail later under the aspect of social policy – is no longer exclusively a "public affair," in which anyone can gain the right of admittance by paying his/her taxes and other social contributions. Social security today means – and we are still in the early stages of development – recognizing a kind of bond and trusteeship with the environment and the generations yet to come. This is a personal trust which can only be formed by making the most of one's own intellectual and social talents. From a perfectly practical point of view the government and its citizens are under obligation to cooperate with one another. It is the business of public authorities to create favourable infrastructural and institutional conditions for the future, if necessary in opposition to initial resistance. Through personal effort the individual, on his part, is responsible to the community for making such investments in the future worthwhile.

The promise information technology holds out for us in the way of more individual elbowroom is not by any means an arbitrary license to do as one chooses. Additional skills encompass taking on new duties. Technology can initiate steps towards overcoming the divided freedom but it is a question of ethical responsibility whether the task will be completed or not. History is offering us perhaps a unique opportunity to make up for past omissions: A purely individualized technology can restore a balanced relationship between the individual and his community. Just as it is not enough to simply want to make use of the rapid strides in science and technology as an instrument of economic development while ignoring their value components, one cannot justify turning our backs on progress, generally leaving others to pay the price for our obstinacy. The general growth in knowledge and information places a demand, not so much on the person of the individual as such, but rather on his personality in relation to the community at large. Only if we are able to recognize this fact, if we are able to integrate and incorporate technological and social progress within the framework of an individual system of ethics, can enlightenment also achieve the full potential of its moral postulate.

The natural sciences, remarkably enough, appear to be especially open-minded with respect to these types of questions concerning moral values. Working intensively with the structural elements of information and communication makes one more aware, or so it would seem, of the importance of the various aspects of development which cannot be ascertained by rational means. The better able machines are to reproduce logical correlations, the more apparent the differences are between computer intelligence and the unique human ability to go beyond this and impart a sense of social ethics and culture to the order of things. The uniqueness of intuition, of empathy, the wide range of interactive thought processes in comparison to mere factual knowledge, our innate need for personal caring – these are all in direct contrast to the algorithmic principles on which computers operate. "Artificial intelligence" is capable of producing specialized knowledge and material goods, but thinking and acting in the human sense of the words is called for when it comes to problems which cannot be solved by following a simple yes or no course of procedure.

The information technology is a revolutionary force on two different levels: First of all by virtue of the dynamic of its development potential and secondly because of the wide-openness of its limits. On the one level it calls on us to use our intelligence for the good of the community at large while at the same time emphasizing the essentiality of the personal element in all our social and cultural relations. We are being faced with a challenge in the new, upcoming technology but this does not imply by any means an inevitable isolation of the individual. On the contrary, it allows for a new form of social freedom in contrast to the collective patterns of order and welfare as developed by the industrial society. In the following we will be analyzing in greater detail what consequences this will have for concrete economic and social conditions in a country like the Federal Republic of Germany.

Chapter 14

Social Benefits at the Expense of Our Future Security

The Crisis of the Welfare Colossus

In terms of the overall structure of the political system one can differentiate in the Federal Republic of Germany between two large system complexes: the market economy and the social welfare state. The market economy arose after the war in conscious opposition to the centrally administered economy of the Third Reich, as the economic equivalent to the newly won freedom. For lack of funds the social welfare state as such had not yet come into its own as an autonomous force; the concept of a "social market economy" (seen in connection with names such as Alexander Rüstow, Friedrich August von Hayek, Wilhelm Röpke, Walter Eucken, Franz Böhm, Alfred Müller-Armack and Ludwig Erhard), functioning as a counterweight between the interests of the individual and society, certainly did exist, however. Additional factors led to the establishment of a "democratic welfare state," embedded in constitutional law and following the pattern of a general system of social security as practiced since the time of Bismarck's reform movement.

Right from the start two developments were established, so to speak, as inherent to the system: The duty of the government to take on the responsibility of distributing funds, in accordance with economic growth, so that the welfare state's deficit could be reduced as rapidly as possible, as laid down by constitutional law and in keeping with sound business ethics; and secondly the structural principle of a "hermaphroditic" social system of insurance laws and regulations, which is "a cross between a social policy of compulsory insurance and a self-governing system of directives" (Peter Koslowski in "Chancen und Grenzen des Sozialstaats" [The Opportunities and Limits of the Welfare State]).

In the following period a dynamic growth in the economy produced a dynamic increase in social benefits, and not just de facto but also de jure. The built-in increases in old age benefits, which were introduced in 1957, can suffice as just one example of the general trend. The expansion of social benefits took two directions: First of all in striving to protect people from as many risks as possible and secondly, within the individual spheres of risk, to provide a maximum of protection. In order to achieve both, not only the range of social benefits to be distributed, but also the administrative structures of the distributing institutions, had to be continually expanded. The welfare state gradually established itself as a state within the state with its own regulations and structures.

The self-propelling mechanism of the welfare state became especially apparent after the inordinate growth in the economy, which had held for around twenty years, began to drop off. Just as the "economic miracle" began to give way to a more normal state of affairs the "social miracle" of the seventies really started to get underway. The welfare state was apparently capable of distributing more than had actually been produced. Obviously Ludwig Erhard was wrong in his foreboding, which he gave voice to as early as 1959, that "no state and no national economy can give away more in the way of social benefits than its citizens have earned."

In truth, of course, even the social policy of the last decade was not able to do away with the basic laws of economy. Quite apart from the fact that there was a lack of political motivation to change things, the welfare state's system of spending was bound up to a course of galloping growth. Without fundamental adjustments being made it no longer even had the mobility to be able to adapt to a slower pace of economic growth. The theory that in a social market economy a part of the economic force could restructure itself as social benefits to make up for hardship and to provide security against existential risks turned out, once and for all, to be simply wishful thinking.

Obscuring this fact, however, was not a difficult matter. One simply had to continue to simulate a dynamic development of revenue intake – by means of inflationary profits based on progressive taxation, by raising contributions little by little, and more recently by the government becoming a regular customer of credit institutions. Keynes' model of preventing economic fluctuation by means of deficit spending was readily referred to in the way of justification – which meant nothing but sinking further into a swamp of self-deception. Even though the economy received a short-term boost, without changing the structure of social benefits the deficit could not be reduced, since, in conformity to the system, higher tax revenues had to be used right away to cover the rising costs in social benefit spending. Accordingly, there was no money left worth talking about to pay off the national debt.

In terms of the overall structure of society a dramatic shift in emphasis between the two large system blocks, the market economy and the welfare state, took place: The social market economy was transformed into a free enterprise welfare state. The rigid structure of social entitlements and benefits, which monopolized a third of the gross national product, dominated the free enterprise system which was geared towards flexibility (and which therefore could not put up much of a fight). The real pressure on maintaining a course of economic growth was exerted, and is still being exerted today, by the demand for keeping up the standard of social benefits, in other words by the spending end and not by the demand for economic goods.

A telling example of what I am talking about is related by Paul Kevenhörster in his book "Politik im elektronischen Zeitalter" (Politics in the Electronic Age). It concerns the work of the data processing team of the Federal Department of Labor and Social Security in connection with the reform of the national health insurance program in 1971. They based their calculations on an annual growth rate in the economy of almost nine per cent. Only by assuming a dynamic growth in social contributions due to increased earnings, could the foreseeable rise in medical costs be covered without having to either increase the contributions as such, or to raise the income cut-off point at which an employee has to enter the national health insurance plan. (In the Federal Republic of Germany all employees are subject to compulsory health insurance unless their earnings amount to more than approximately 35,000 dollars annually. trans. note) "Raising the cut-off point would have met with strong political opposition," Kervenhörster admits. "Those responsible chose to declare the first assumption of a strong growth in the economy to be most likely and neglected to make any other plans in line with alternative calculations. Otherwise measures for cutting back on the exorbitant medical costs might have been initiated then instead of waiting till later. A monkey-see-no-evil attitude was preferred, however. After all, high growth rates had become a matter of course."

The free enterprise welfare state shows unmistakable signs of developing into a colossus: It is an organization, but an omnivorous one. Its need for money chronically exceeds its economic means so that it either goes deeper in debt or is forced to maintain a spiral of growth which exceeds actual consumer buying needs. The next generation will thus be doubly handicapped: Without the burden of these old debts, simply on the basis of their own economic endeavours, they would be able to enjoy a higher standard of living and a healthier, better environment. The crisis of the welfare colossus will become a society crisis of credibility. A society which pursues a policy of handing out social benefits at the expense of the next generation's security, can hardly claim to be future orientated. The here and now clearly takes precedence over the hereafter.

Most social security plans are designed accordingly: Married couples who both work and have decided not to have children can afford more than the average in the way of material goods and end up acquiring generous social security benefits. People no longer feel compelled to make provisions for their old age and the steady increases in social security contributions places an enormous burden on the next generation because there is a smaller workforce supporting a larger number of senior citizens who are legally entitled to liberal old age benefits. The recent reform of social security regulations, which takes the years spent bringing up children into account, is a step in the right direction towards doing away with the blatant discrimination against non-working mothers. Nevertheless, it is still much more profitable not to make a personal contribution towards securing our future by bringing up children. As long as this continues to be the case, the collective social security system will inevitably create friction between old and young in contrast to its own self-image as an ideal form of cooperation between the generations.

If the ability of the system to provide security for the future does not look too promising, one might at least expect that it is doing an excellent job of managing the present distribution of benefits. Obviously there are also shortcomings in the system here, though. For just as the welfare state does not reward doing without things in order to make provision for the future, it places a premium on not taking advantage of benefits. A large slice of everyone's paycheck goes towards social contributions and it is irrational not to avail yourself of supplementary social benefits and services, since you end up paying for them in any event. The savings resulting from individual economizing are distributed among everyone, are collectivized, whereas additional consumption benefits the individual directly. Those who try to take advantage of as many services as possible, in return for the contributions they have been forced to hand over, end up on top. Since people are left completely in the dark about the actual costs of using social services, they do not even have to feel guilty about taking advantage of them; no one is there to destroy the illusion that they are simply pocketing with their left hand what they have already paid for with their right hand. As a result, the costs of social services continue to grow at an inflationary rate, forcing at some point an increase in the amount of contributions, which in turn sets the next rotation of the social benefit merry-go-round in motion.

What is right for the consumers of social services is only proper for those providing such services. The principle of reimbursing costs in full, which still holds, does not make economizing an appealing alternative. Hospitals and senior citizen homes which try to reduce costs cannot expect any kind of recompensation. On the contrary, they risk having their hospital and nursing charges being cut back. A sensible economic course, namely to save, is nevertheless not rational.

The existing social system is thus faithfully mirrored by the way in which its institutions function. A redistribution apparatus, which by now has around 250 billion dollars annually to distribute at various levels and in a wide variety of classifications in a kind of circulatory administration, inevitably produces bloated and autonomous structures. Thinking in terms of tables and drawing up forecasts of quantitative needs are the order of the day. The more plausible one's requirements seem the greater the chance of receiving financial support. Under these circumstances rationalization measures do not primarily affect either the organizations (considered necessary for coping with the demand placed on them by the sheer volume of output) or new equipment budgets (it is usually an easy matter of justifying such expenditures to the non-experts who are in charge of granting approval), but rather the expensive personal services.

A welfare system that shows clear indications of dividing people against one another, of reversing values, and narrowing future options – is this really still in the spirit of the constitution and the original concept of a social market economy? Does the division of responsibility between the individual and the collective still do justice to the need of protecting the weak? Or is it not rather a case of an undifferentiated, ongoing egalitarianism which does not give the really needy enough while those who are only along for the ride are given much too much? Can the negative consequences of a system which has grown into a colossus – crippling social mobility and individual initiative, with an uncontrollable momentum of its own and wasting resources by uncoordinated subdivisions – be managed in the long run or be successfully combated alone by selective budget saving measures?

When the whole thing is no longer consistent in itself and is a muddle of tangled knots, there is no sense in just pulling at one end. We have to reconsider our system of values, the political imperatives, and the means we employ.

Chapter 15
Rediscovering the Human Factor

In Defense of a New Social and Economic Policy for Modern Society

The classical antithesis of capital and labor – the "worker problem" of the 19th century, which was the starting point for a social policy of differentiating between the economy and the welfare state – no longer exists today in this form. On the one hand manufacturing capital has become "impersonalized" to a large extent. Large family owned companies have become rare, joint-stock companies with a widespread distribution of wealth are the rule from a certain size up. The rights of union representatives in having a say about things – in factory committees, on the board of supervisors and on the board of directors – extend not only to personnel and social matters but also include important business concerns. Social plans and regulations regarding protection against wrongful dismissal as well as the more or less generally binding nature of collective bargaining agreements rule out arbitrary acts on the part of companies – so much so, in fact, that the only legal "arbitrary act" open to a company owner, hiring employees, is only ventured upon after careful consideration.

On the other hand many private households have acquired a considerable amount of valuables and capital (the latter above all in the form of benefit rights). West Germany's citizens are without a doubt among the wealthiest of the world; their incomes as a rule allow them to acquire considerable savings or to fulfill fairly luxurious consumer wishes.

The "social problem," in other words, is no longer connected to the distribution structure of manufacturing capital as was still the case at the beginning of the century. Today there are "employee capitalists" who manage their assets with business professionalism as well as "social employers" whose dealings are strongly characterized by social ethics. There are – legal or illegal – part-time entrepreneurs who know how to lucratively combine dependent and independent work as well as small tradespeople, who would end up with less than they could get if they lived on welfare, if they did not "exploit" members of their family.

The social battlefield runs along different lines nowadays: between those who have jobs and the unemployed, between senior citizens entitled to generous social security benefits and pensions and those who have to get by on the minimum rate of benefits, between childless couples and families with a large number of children, between "clever" and "stupid" consumers of the welfare state.

In view of this situation, it no longer seems justifiable to cling to a dualistic block system which is no longer corroborated by large social counterpoles ("capital versus labor" or "free enterprise versus social justice") and at most is mirrored in the rigidity of society's institutions. The increase in the earning ability of many citizens must find its counterpart in the gradual and politically carefully thought out supplementation of public benefits with elements of private provision. The principle of giving everyone a slice of the cake has to gradually be replaced by a welfare state which is orientated towards helping specific target groups. The fact that only ten per cent of social services are paid for privately while 90 per cent are provided by public or semi-public sponsors (thus Joseph Huber in his contribution "Duale Sozialpolitik – Fremdversorgung und Eigenbeteiligung" (Dual Social Policy – Public Benefits and Private Contributions) to the previously mentioned book "Chancen und Grenzen des Sozialstaats" (The Opportunities and Limits of the Welfare State)) means that many well-earning citizens are getting services that they could easily afford to pay more for themselves.

Since financing the welfare state on the other hand involves heavy social contributions and a progressive system of wage and income taxation, many normal wage-earners are being taxed beyond their limits. A reorganization of the taxation system is without a doubt the most effective countermeasure that can be taken in the long-term – reducing income taxes and increasing sales taxes. At the same time an employee's private contributions towards retirement funds and pension schemes must be strongly encouraged. One should not, however, have any illusions about the period of time it will take to make such adjustments. This is why it is all the more important in the meantime to establish self-help and privately funded systems on a small scale which could slowly but surely put a check on the claim to sole rights by public interest factions.

Social wards and private social initiatives, which have recently cropped up in increasing numbers, indicate the right direction to take. The government, however, should not suddenly discover the principle of subsidiary systems in this area, which it has reneged again and again within the framework of its own public social institutions, and use it as an excuse for not giving adequate financial support. Success particularly depends here on a policy of "active subordination" (Warnfried Dettling), which recognizes that decentralized self-help with comparatively modest means can often accomplish things more effectively than a professional social institution which is geared towards total maintenance.

It is interesting to see how one, from quite different analytical approaches, keeps arriving at the same conclusion: the need for a social order with smaller units which incorporates more self-responsibility and more self-provision. The historical background is formed by the changes in economic conditions which have resulted in overcoming the old social problems. The fiscal cause is the financial strain on the government by large-scale social structures (which are used most intensively and are therefore most expensive when income tax earnings are at a low ebb, namely in times of a stagnating economy). From the standpoint of legal and ethical values and in view of the technological development, though, one also keeps arriving at the same conclusion: the need for affirmative action. The parallel between these different movements, which at first glance do not seem related to one another, is almost amazing and certainly justifies talking about a historical need for a new policy of social security.

For our constitution, for example, as the Federal Constitutional Court emphasized as early as 1954, " the image of man ... is not one of an isolated independent individual; the constitution has rather defined relations between the individual and the community in such a way that man's ties with the community and his integration were stressed without, however, infringing on his intrinsic value as a human being." The founding fathers of German democracy wanted by these means to take two teachings of the last two centuries to heart: that formal freedom rights, which only consist of anarchistic rights in opposition to the government, can nevertheless produce conditions beneath human dignity if there is a lack of minimum material and social security; on the other hand the social order should not be allowed to degenerate into a collectivized organization which sees the individual as simply the object of impersonal decision making processes. The conflict between individual freedom and responsibility to the community, according to the will of the constitution, should be resolved at an "intermediate level," between the extremes of isolated independence and collectivized decision making.

The separate social benefit laws, taken individually, are certainly on firm constitutional ground. Can this be claimed with the same assurance, however, for the sum of the different forms which the welfare state has taken on by now? A basic legal order only really makes sense if, not just its individual elements, but the sum of its parts as well reflects the reality of the constitution. The reality of the welfare state, while certainly unintentionally, is that personal social activities are impeded and prevented more often than they are given encouragement. Public welfare and care-taking by anonymous organizations are given priority over the classical principle of helping people to help themselves. By making it so profitable to take advantage of social benefits, the idea of man being bound up with his community has ended up being perverted into an open invitation to put burdens on the community.

To what extent does the principle of decentralized social responsibility which, in theory at least, is an element of our basic law to the extent that the constitution feels called upon to caution against superfluous restrictions on the part of collective authorities, still play a role in practice? What remains of the principle of solidarity if a real sense of community spirit is no longer possible, if the government has standardized and monetarized the social relations of its citizens to such an extent that personal service and philanthropy almost seem to be adversaries of the system? And, finally, to what extent does the principle of equality still exist in reality when an overabundance of social slogans has made the difference between the needy and non-needy almost completely unrecognizable?

It is about time that we start seeing the welfare state not just as a financial and organizational problem, but also carefully examine it for discrepancies which are at variance with constitutional law. This cannot be accomplished, though, without openly facing its current lack of values and without a clear idea about the political alternatives.

The more society has relieved people of the duty of individual responsibility by endeavouring to establish a perfect system of social security, the less capable people have become of forming conflict free relationships with those close to them – in the family, with neighbours, friends and colleagues. Our ambitious welfare society with its legal guarantees has a hard time in keeping track of and organizing matters in the area of interpersonal relationships.

Examples can be found everywhere you look: Psychologists rate weekends and holidays as especially critical times for familiy life; neighbours who used to help each other in building their houses now file protest suits against the building petitions of the other just to be on the safe side or sue them for violation of property line rights; children go to court against their parents concerning the amount of their allowance and for how long they are entitled to support. The individual has learned to successfully fight large-scale organizations by taking full advantage of every available right, looking upon money payments as a way of compensating for a neglect of duties or responsibilities. Involuntarily, this formula is applied to personal relationships. One's own point of view is the only bona fide one whereas the other fellow's interests are only given formal credence to the extent that they can be upheld by a court of law.

"Giant organizations are independent of man and nature but also alienated from them. The systems are deficient in interpersonal relationships and communication between people no longer functions properly," Philipp Herder-Dorneich writes in his essay "Ordnungstheorie des Sozialstaates" (The Theory of the Welfare State). "Providing personal care for others in time of need used to be done as a matter of course but nowadays it is something that has to be consciously organized. How can we meet this new challenge in terms of a new social policy? Organizing primary groups, not just secondary systems, is a job that policy makers are unfamiliar with. Indeed, it is not my intention here to plead for a new social policy."

A new form of social policy at the level of interpersonal contact is needed, not just to loosen up the rigid structure of the welfare state, to overcome financial bottlenecks, to bring the discussion of our basic system of values out into the open again, and to gradually lay the groundwork for social and taxation policies which are in harmony with one another after a century of conflict. The political rediscovery of the human factor, the rehabilitation of primary virtues such as helpfulness, charity, and sharing responsibility, is rather a very social mandate which not only the government owes its cititizens but which the citizens also owe themselves and the community.

The time for action has come *now*, and not just because we have run out of money. People are becoming increasingly emphatic in voicing their need for a new change of direction. We are also undergoing a structural change in technology which holds in store the great opportunity of rectifying the deficiencies in interpersonal relationships. We would have to be blind not to take full advantage of the opportunity for reshaping our social structure that the new technology is holding out to us. Attempting to restructure production and labor in keeping with the new information technology while leaving the social system of the industrial society, with all its material and immaterial shortcomings, untouched would be an act of cultural schizophrenia which could end up destroying the community.

A "new social policy" implies therefore something quite fundamentally different from simply a redistribution of wealth. It must be understood as an integrated economic, social, educational, and technological policy on the basis of decentralized responsibility and solidarity. It calls for human technological options in the tradition of a social policy which sees itself as a moral science. It creates a framework in which primary relationships can be indirectly strengthened. It takes into consideration relevant aspects of educational policy, environmental protection, cultural events. Under all this – not on top of it! – there is a safety net of social services to provide protection for all those who are really in need of it.

In this sense of things the decreasing importance of human labor in the manufacturing of goods is not a disaster but rather a piece of good fortune in that it contributes towards speeding up the process of establishing a more differentiated private service economy. Once flexi-time has been established on a broad scale, taking individual wishes into consideration as much as possible, more and more people will be able to take, not just days, but weeks, or even months off work all together. To use all this free time exclusively for recreational and consumer purposes would not only be beyond the means of most people, it would also be seen as an empty, meaningless way of simply killing time. The desire to play the part of a "producer" oneself would inevitably arise but not everyone is willing or able, according to his/her inclinations and talents, to be an enthusiastic do-it-yourselfer or hobby gardener. Many people would consciously want to look for work in which they could be of service to the community and which would provide them with personal contact and give them the satisfying feeling that they are doing others good.

The basic structure for integrating this new type of service work already exists; it simply has to be developed and enlarged in a satisfactory manner.

Out-patient Services as a Case in Point:

In Baden-Württemberg – I am using my own state as an example, not because I feel that conditions here are especially exemplary, but simply because I am most familiar with them – there is a total of 364 out-patient services. Currently 2,600 full-time and 1,500 part-time trained employees, 5,000 voluntary aids and 140 conscientious objectors (In the Federal Republic of Germany young men who are exempt from military service on the grounds of conscientious objections are under obligation to do alternative civil service. trans. note) work in these services. In 1983 these out-patient stations were able to provide the following services: looking after the welfare of 170,900 infirm and aged people, around nine million house calls on 117,000 workdays, out-patient treatment of 157,000 people, and conducting 5,500 first-aid courses. The whole complex costs the state of Baden-Württemberg 12.5 million dollars annually, all of which goes towards personnel costs – about 4,150 dollars a year per trained employee. In comparison, 12.5 million dollars is what 250 beds in an intensive care ward of a hospital cost in a year.

The feeling of those who were able to take advantage of such services and were saved from being prematurely pushed into an old people's or nursing home, knowing that their social security payments would not be enough to cover the costs, cannot be weighed in dollars and cents. One cannot measure either how much loneliness was averted by such house calls. It does not require a lot of imagination, though, to realize just how much more can be accomplished, in addition to what is already being done, and that this in turn depends directly on an increase in the number of voluntary and part-time aids.

Musical Education as a Case in Point:

More than 200 music schools have been established in Baden-Württemberg in the last few years in which around 2,000 music teachers are employed full-time. Pupil's fees cover half the costs of these institutions, more than 35 million dollars annually. This indicates not just a marked willingness on the part of parents to use a part of their available income for the purpose of privately educating their children instead of just spending it on more and more consumer goods. It also indicates the basic pattern for a new division of labor between the secondary and tertiary sector: Increases in productivity due to machines in the one area can very well, by means of better earnings, form the basis for creating new jobs in quite different fields of activity.

Adult Education as a Case in Point:

Every fourth adult participated in an adult education course in 1982; by now the proportion of such participants has probably risen even further. Not just job orientated training programs and courses are enjoying an increasing popularity but the demand for arts and crafts courses is also growing. The Adult Education Commission of Baden-Württemberg announced in their report to the government at the end of 1984, for example, that in the previous year the Adult Education Centers (non-affiliated public evening schools, trans. note) alone had conducted almost 13,500 courses with 168,000 participants in the manual skills and in field of arts, in contrast to 2,200 courses with 37,600 participants ten years previous to this. There was a similar development of interest in courses dealing with health education.

The "fourth education sector," which by now is quite an appropriate term for adult education, is still not state regulated to any great extent. It is organized by a variety of private sponsors (companies, professional associations, chambers of commerce, unions, academies, Adult Education Centers, religious organizations, private institutes, and schools) and offers a broad range of courses and activities with individual options. Within the next couple of years certain special interest groups will be wanting to take advantage of the private education market's offerings in increasing numbers: senior citizens – the number of people over the age of sixty will have increased by around a million by the year 2000 –, who want to make their third phase in life as active and meaningful as possible; women, who after raising their children, want to try to get back into the job market; unemployed men and women, who want to increase their chances of getting a job and alleviate the psychological pressures of being out of work; working people, who are interested in a creative use of their free time to compensate for stress on the job.

In the future one will increasingly encounter the direct and indirect influence of the new technologies in this area as well. They will be important, first of all, as a means of increasing productivity, creating the necessary "service buying power." Secondly, the adult generation in particular will be confronted by the need to learn how to deal with them. Learning about computers, what our children can pick up at school and through their use of consumer electronics without much of an effort, is something the older generation will have to consciously work for. Not letting the information and know-how gap between the generations get too wide, calls for an extensive service sector with maximum technological, social, and didactic qualifications.

What we have already found to be true from an individual ethical point of view is clearly substantiated in all these fields: The new technologies mean an enrichment of our intellectual capabilities but at the same time they also incorporate an ethical challenge. We fail to live up to this challenge if we try to reduce the complexity of the economic, social, and cultural interrelationships to an oversimplified dogma of distribution in accordance with anachronistic patterns of order. The fall of the welfare state is the rise of the social society – this is the doctrine of a new social policy.

How can we break new ground, though, if the old, self-inflated welfare colossus apparently has extended its tentacles into every hole and corner? Certainly not with "fire and sword," no matter how tempting the idea might at first appear in view of our empty coffers and rigid structures. A social demolition derby is neither politically feasible, necessary, or even desirable. What we rather need is a social reorganization with the coordinated reform of the various component systems under the aspect of an integral social policy.

In summary this calls for a number of changes:

First of all, the change in economic conditions, overcoming the old social problems, should find its echo in mobilizing people to take more economic initiative. Helping new businesses get started also directly influences social policy in that it lays the foundation for more self-reliance in the long term. Being able to give growth new momentum does not necessarily depend on either the production or the distribution aspect of the social market economy.

Secondly, the conversion towards a private service economy calls for active support in the form of government subsidies. Individual initiatives, self-help groups, and non-government forms of organizations should be encouraged in their efforts toward relieving the strain on our social system.

Thirdly, the government should gradually but steadily pare down its social benefits. This includes, not only cutting back on superfluous hospital beds, but also reducing the amount of subsidies and material transfer benefits that are no longer justifiable, ideally by decreasing the amount of subsidies by a certain percentage every year.

Fourthly, increased transparency and equivalence should be made the basis of criteria for judging public social service systems. Cost/return calculations, drawn up by the administrative bodies of the social service institutions, should be submitted to committees in charge of granting financial subsidies. The individual who uses the available social services should also be aware of just how much they cost.

The next logical step would then be to replace the compulsory full-coverage system with more flexible insurance programs that allow for more freedom of choice. Benefit payments could be worked out and vary according to the amount one had privately contributed. Protection against fundamental risks would remain untouched – the government by rights must act as a trustee in this regard. As far as the various supplementary benefits go though, there is no justification for either the government having a monopoly on such benefits or for individuals being forced to enter such programs.

To put it into direct terms, it is a mistake to think that everyone expects and wishes to live out their old age at the expense of welfare in old people's or nursing homes. Many people would be happier if they could pay for being taken good care of (both socially and medically) in their own homes with the benefit payments from various insurance plans and retirement funds that they had been able to invest in while still working.

Fifthly, social policy should contribute towards strengthening primary relationships. There is a variety of different ways in which this can be accomplished: tax deductions for behaviour which is of particular social value (such as taking care of infirm relatives), the extension of cultural and social infrastructures on a small scale by means of complementary financing (granting public subsidies to private projects that have proved themselves capable of taking initiative on their own), a reduction in insurance payments and tax deductions for community activities, to name just a few examples. If, on the opposite side of the pole, going to court was made somewhat more expensive, many a conflict of interest could be settled over the backyard fence instead of being hashed out in a court of law.

None of these are long-range plans, but are simply steps in the right direction. Far-reaching reforms, especially in the area of taxation and social security regulations, are in store for us during the nineties. These reforms will require many years of careful preparation by experts. The way has to be paved, though, by a change in our mental attitude, a new understanding of what the freedom with social responsibility is all about.

Chapter 16
A New Look at an Old Problem

The Economy and the Government

Not just social reality has undergone a radical change since 1949; the development of the economy has also taken off in directions which do not correspond to established economic theories. This has been the case for decades and has rarely been the cause of any controversy. The combination of the change in our economic structure and high unemployment, however, has placed the question of the government's role in determining industrial policy unexpectedly in the foreground.

A lot of what has been written and said concerning this question is worth thinking about in regard to the possibilities and limits of private and public actions. A great deal, though, falls under the category of "religious battles," for which, if anyone at all, the politicians in any event should not feel responsible. One should also take care not to fall prey to any illusions in judging economic patterns of development from a social policy point of view: Not everything which brings prosperity is necessarily the result of a culture of free enterprise. Not everthing which gives rise to crises is indicative of political incompetence or stubbornness on the part of the unions. The social market economy also sees itself in its economic components as a "principle of design" and an "order of peace" (Müller-Armack); it delegates the responsibility for upswing and recession, for prosperity and stagnation among everyone.

The post-war success of the social market economy was an impressive confirmation of a conclusive concept which was able to be realized at the right moment. It is not slighting this splendid accomplishment in the least to add that without a number of special circumstances, which had nothing to do directly with economic criteria, the German economic miracle would have turned out to be on a much smaller scale or would not have taken off as rapidly as it did. Among these circumstances the following deserve to be mentioned:

- the motivating force of making a new beginning in the newly won freedom which went far beyond the normal, profit-orientated business interests of established economies;

- the enormous discrepancy, as a result of the war, between the amount of goods available and the demand for them which contributed tremendously towards speeding up the establishment of production capacities;
- the opportunity, in the wake of the reindustrialization of the ruined economic structures, to make direct use of the technological knowledge of management and the know-how of a quailified skilled labor force in establishing modern forms of production;
- a functioning system of domestic markets, which were still self-contained for the most part, and which, moreover, had the advantage in the early years of a trade policy of solidarity by the West which was atypical in terms of the usual rules of economic competition;
- the generous, at the time essential, capital investment help of the Marshall Plan by the USA.

It is not just of historical interest to recall to mind these remarkable circumstances which gave birth to our unprecedented success. What is more important is that they explain why the economic sins committed in the reconstruction period did not have any noticeable negative consequences at first. A dynamically growing economy can afford to put up with a number of adverse circumstances, even some which are at variance with the system itself. It is only when there is less distribution leeway, which is the usual case in times of normal economic growth, that mistakes in judgement become painfully apparent.

Since the war, for example, the building industry has hardly ever been subject to the laws of economy. It was not until 1961, through the Lücke Plan, that its close connections to a central administration of economic planning were severed. Even then it was one of the first areas of industry to receive large amounts of government subsidies and it has consumed more than 100 billion dollars in tax money. Despite this Alfred Müller-Armack, among a number of others, has argued that the building industry subsidies, rent assistance, and tax deductions are measures that are in line with a policy of sound economics.

It would be more appropriate to call them by another name: They are useful methods for dealing with a state of emergency. There can be no doubt, however, that their basic character is one of government intervention. Just how far out of touch they are with the reality of the market can be seen in the results of the apartmenthouse building industry alone. Public housing projects continue to exist, although a third of the people living in them are not entitled to such housing, and they have to be subsidized by as much as 7,500 dollars a year per living unit, while at the same time one and a half million people receive rent assistance because they are living in apartments, they have found on the open market or public housing ones, which are beyond their means. This is an absurdity that has been produced not by applying the principles of sound economics but by their rape.

The socio-political ambitions of the guiding spirits of the social market economy, their conviction of not only being capable of but also under obligation to compensate for the "blindness" of the competitive process by a system of social ethics, is not to be critisized by any means. It has resulted again and again, though, in the government taking measures for the redistribution of wealth at the source of competition, in the production of goods.

Finally – as the example of the regulated housing market also indicates –, there has more or less always been a consensus that there are areas of the economy that must be given special public consideration and in which undesirable circumstances of the market should be justifiably compensated for by means of government countermeasures. The influence of this consensus is most apparent in the so-called exceptions to the laws against the restraint of competition. Among others, the coal and steel industries, agriculture, utility companies, and public transportation firms are for the most part not subject to regulations safeguarding competition. "In cases where the irregularities of the market are not wanted due to police, political, or military reasons (for example, in order to ensure that utilities continue to function in times of emergency or crisis) measures for maintaining structures should be made as transparent and as efficient as possible," as Egon Tuchtfeldt puts it in his contribution, "Stilelement der sozialen Marktwirtschaft" (The Element of Design in the Social Market Economy), to "Grundtexte zur sozialen Marktwirtschaft" (Essays on the Social Market Economy) published by the Ludwig-Erhard Foundation. Political countermeasures for the purpose of maintaining economic structures – these are, to put it in plain English, nothing other than maintenance subsidies.

As early as 1966 every job in the mining industry was subsidized by 467 dollars (fourteen years later it had already risen to 5,500 dollars worth of government subsidies per miner). At that time the steel industry only required four dollars of government subsidies per employee (in comparison to 127 dollars at the beginning of this decade). The conviction, though, that steel formed an indispensable production basis for the rest of the economy was so widespread, until recently, that the economic justification for government support measures could be made without practically anyone raising an objection. In Wilhelm Fucks' book, "Formeln zur Macht" (1965) (Formulas For Power), the "steel religion" of a whole era is strikingly depicted. It almost appears grotesque when one compares the projections made at the time for national steel production, and the political power one expected to achieve in connection with it, and the conditions that nowadays prevail in the steel branch.

For similar reasons, as part of the Treaty of Rome negotiations in 1956, it was considered in the final vote as politically justifiable to exclude the agricultural sector from the market mechanism of a domestic economy and to subordinate it to a European bureaucracy. After all, one was dealing here with an area of "supply security" in which government regulations for the purpose of achieving higher aims did not appear all too offbase, even from a free enterprise point of view. The chairman of the "Common Market" committee in the government negotiations 1956/1957 in Brussels, Hans von der Groeben, wrote in retrospect in 1982: "The original concept for agricultural policy was based to a very large extent on the principles of free enterprise. In the course of the negotiations by the Council of Ministers one deviated more and more from these principles until the regulations which were finally agreed upon opened the way for the dirigism and protectionism which were later practiced."

That this policy crime had nothing to do with considerations of material structure, but was simply the political price that had to be paid to France to get her to agree to the EEC, indicates the range of possible reasons one can come up with if one has once determined to sanction government intervention in structures on the basis of free enterprise. From today's standpoint there is no denying that if people had stood by their principles, Europeans would have been saved a lot of problems and expense; to be fair, however, one should add that Germany's negotiating position at that time, twelve years after the war, was not as strong as it is today.

People are continually emphasizing the government's being justified in offering "public goods" in those areas where the economy is not in a position to satisfy the total demand. This is basically nothing more than a concession of theory to the need for changes in practical policy. Investments in highways and education were naturally regarded as matters of public priority in the years of reconstruction; as a result they were included in the catalogue of infrastructure measures for which the government felt itself to be responsible. Ever since the environment has become a matter of political priority, the textbooks all remind the government of its duty to create a healthly environment or to mend the ecological damages. This kind of direct demand for action obscures rather than supports the primary responsibility and guiding function of the government.

Mind you, arguments can be found for all these measures, but they are pragmatic arguments and not predetermined political axioms of a social market economy. If this were the case, politicians would have to start asking themselves if investments in education are always really a "public good" or if they have not actually often also proved to be a "public waste;" whether building new highways can still be regarded as a legitimate rumpus room for an enormous bureaucratic apparatus in view of the discrepancy between countless drawing board plans and the relatively small number of projects that have been realized; whether automatically relying on the "state" to be the nation's environment protector does not tend to weaken rather than strengthen the innovative force of the industries to achieve an edge in competition by developing products friendly to the environment on their own initiative.

To sum things up, there is no way of avoiding the fact that the role of the government in our country has never been limited to the classical free enterprise pattern of delegating authority – establishing a framework in which competition, business, and foreign trade policy can be made while being reserved when it comes to a social and fiscal policy of distributing wealth. This is clearly proved by the federal budget expenditures of almost 15 billion dollars in subsidies per year (the total amount is probably around 40 billion dollars), by at least 750 billion dollars worth of state-owned businesses, by the sprawling welfare system, and last, but not least, by the omnipresent discrepancy between the theory and practice of free enterprise.

In view of this situation, there is no sense in making the difficult phase of economic restructuring even more complicated by arguing about principles whose model character no one seriously questions. At the same time, there should be no illusions as to the compromises and shortcomings involved in dealing with them on a day to day basis. It seems much more important to me to examine the question as to what conclusions will have to be drawn from the radical changes being made in national and international economic patterns in view of an effective policy of free enterprise.

We have already discussed a number of concrete changes that the new technologies will incorporate. In this connection three aspects of the structural change are particularly of interest and are especially important in terms of policy: the change in production basis, the new definition of the term infrastructure, and the questions as to the "liability for values" of the market.

A policy of free enterprise, first of all, should be fully aware of the fact that the production basis of highly industrialized countries is undergoing a radical change. Information and communication no longer simply play a role as special forms of technological application (as one believed in Europe until just recently), but have become production factors in their own right just as labor, capital, and real estate have always been.

A process of suppression as well as of change is taking place which shows parallels to the industrialization movement: The emergence of industrialization gave the primary sector as a producer of goods no longer the same economic importance, while at the same time the use of new technologies was able to increase productivity in leaps and bounds. Similarly, the secondary sector, especially as far as mass production goes, will no longer function as such a dominant factor in economic and employment policy while the use of information technology, however, will continue to result in increased productivity.

Stated in even more general terms: "Every increase in productivity in the history of man was the result of information processing, in that innovation (in the sense of Schumpeter's use of the term) is basically the effective use of scarce production factors in order to cope with changing needs. One could also say that from time immemorial information activities have been rationalization measures for other forms of activities ... As information processing rationalizes itself, in that it has developed a technology which can increase its effectiveness, one can expect a corresponding increase in the total economic development of productivity" (Christian Lutz in "Westeuropa auf dem Weg in die Informationsgesellschaft" [Western Europe on the Way to Becoming an Information Society]).

The objection one often hears, that the government has no business in structuring the policy of a specific branch of industry, ignores the fact that we are dealing here with much more than the simple question what direction certain lines of technological development should take. Micro-electronics, computer technology, and sensorics, but also the innovations in the development of new materials and production processes, are no longer simply a matter of specific products. For most branches of the economy they have become important as raw material. Nowadays, without integrated information technology, elements for purposes of control, surveillance, and ease of use, industrial machines, household appliances and cars cannot maintain a competitive position on the market.

It has always been considered the responsibility of the government to safeguard the basis of production, especially in bottleneck areas. Foreign trade regulations, usually in the form of direct bilateral agreements, ensure that the raw materials which a country itself does not have but needs are imported. Capital is made available through the activities of issuing banks in keeping with international finance and monetary policy. When there was a lack of labor, the policy of inviting foreign guestworkers to come and work in West Germany was established. There is nothing new in terms of policy, therefore, about the government making it its concern to reduce the "supply deficits" in the production factors of information and communication.

Against the background of a marked middle-class economic structure, two main points of focus are indicated: acting as a mediator in imparting the necessary technological know-how and financing innovations.

As far as the former goes, one should take into consideration that the government itself, as the sponsor of the universities and other public research institutions, already has at its disposal the authority to organize a share in the production of economically relevant knowledge. By not making this knowledge, whose production is under its direct control, available in an adequate degree to small and middle-sized businesses, the government indirectly influences free enterprise competition. There is a tendency, namely, to automatically give preference to large companies which have adequate research and development capacities at their own disposal, which have the information raw material on the premises, so to speak, or which can afford to buy from the universities what they lack in knowledge by granting research fellowships. This would end up encouraging the tendency towards a concentration of the market by establishing a kind of technological cartel; antitrust laws, however, would place the very same government under the obligation to prevent this kind of concentration which endangers economic competition.

This consideration alone indicates just how absurd the arguments against the transfer of technology between the universities and the great majority of our economy (middle-sized businesses) are. Furthermore, the only alternative to establishing direct information and advisory contacts, launching technologically orientated sponsor programs, is even more hazardous from a political point of view. The possibility cannot be ruled out that the administrative authorities responsible for reviewing and granting funds would tend to judge projects on the basis of their hypothetical chances on the market, which does not lie within the sphere of their jurisdiction. There would also be no way in many cases of avoiding giving preference to large companies in granting government research and development funds. It is much more in keeping with the spirit of free enterprise to create structures in which everyone has equal access to the available information and to leave the question of taking advantage of opportunities up to the initiative of the individual entrepreneurs, rather than to carry out a preliminary process of selection.

The organization of information transfer would no longer be a question of principle for the state, but would depend on concrete regional conditions (the structure of the economy, the transportation network, existing educational institutions). In rural areas the chambers of commerce and professional associations would naturally play a more important role as mediators between businesses and public research infrastructures than in the industrial centers. It is quite possible though, as experience in the Black Forest region has shown, to establish direct contact, especially between technical universities and the companies of an area, from which both sides can profit a great deal. The "Furtwangen Model," which consists of a strictly defined transfer of personnel between a university and business, is not only a big help for a lot of companies (particularly in converting from a mechanical to an electronic form of production), but it also gives students and teachers valuable experience in applying their theoretical knowledge.

At the universities and in institutes the production factor knowledge lies, so to speak, directly on the surface, similar to natural resources which can be extracted by open-cast mining. To extend the metaphor, it would be pretty foolish, economically speaking, not to allow businesses to utilize and exploit these natural resources on the spot instead of artificially holding them off at a distance and interposing a government administered system for mining and transporting the raw materials. Exactly this kind of thinking, though, is reflected in many of the arguments objecting to innovation centers and technological factories being built on or near college campuses. If information and communication are authentic production bases, which no one can seriously doubt any longer, access to them has to be regulated by economic criteria and not by traditional role concepts. Otherwise, there is a danger of policy becoming an exercise in power simply for the sake of power.

The financing of innovations depends directly on the availability of venture capital, whether one's own or someone else's. Young business people, in particular, end up going bust in the difficult early years more often for lack of capital than because of poor products.

There are basically two approaches for strengthening innovation and investments in businesses: One can either improve the capital base of companies by lowering corporation profit and capital asset taxes, or one can increase the efficiency of the money market in order to be able to supply the companies with more outside funds.

Understandably enough, the first method is much more popular. Without a doubt, the current tax laws in Germany do not differentiate enough between the lifespans of investments. There are also peak loads as a result of the accumulation of different types of taxation which are extremely high in comparison with the rest of the world. Nonetheless, one has to warn about "half-cocked" tax policy measures. There is the danger namely, that this kind of patchwork, as opposed to a general and comprehensive tax reform, will not result in the hoped for economic benefits. On the one hand, such short-sighted measures usually do not relieve the financial strain on the companies to the extent that they end up producing additional investments, while on the other hand the government's limited revenues are prematurely exhausted.

Raising the rate of degressive depreciation by ten percentage points alone, from 30 to 40 per cent, would lead to the loss of almost three billion dollars in tax revenues within a year. Assuming that the federal government decided, in an act of brute force, to try and make up for the loss by cancelling all the coal, steel, and ship-building subsidies, there would still be one billion dollars to be accounted for. These kinds of tax policy measures, as examples in other countries have also proven, run out of steam very quickly when nothing is done to change the shortcomings and failings in the other structures of the economy.

Incidentally, one should not lose sight of the fact that the German economy has profited from reduced taxes since 1983 which amounted to about 5 billion dollars by the end of 1985. According to the 1984 company report of the Deutsche Bundesbank, the net capital invested has risen from eight per cent (1982) to eleven per cent (1984).

Additionally, isolated regulations which go beyond the Tax Cut Law 1986/1988 involve the danger of not having much of a psychological effect and of narrowing the scope of action for the unavoidable "structural" tax reform. The concept of an authentic tax reform has to be based on the following two principles:

- Simplifying the tax system is of highest priority. Tax law has to make more use of the possibility of standardizing and estimating in a lump sum.
- the only realistic course is a graduated plan, structured in terms of time and function, in which tax cuts are tied-in with a reduction in special privileges and financial aid.

Within the framework of these guidelines the corporation tax reform should pay special attention to the following aims: relieving the tax burden of small and middle-sized companies, reducing the disparities with which the different forms of financing are treated, further reducing the depreciation periods, and bringing corporation taxes in line with other European countries.

The restructuring of the income tax laws also calls for a number of essential changes: The unjustified, sharp increase in the tax rates in the first third of the progressive zone must gradually be converted into a linear progressive rate (cost: 12.5 billion dollars!); the unbalanced distribution of burdens between tax and social legislation has to be amended (a family without means and two children receives 12,750 dollars a year in welfare benefit payments, whereas a working family with the same number of children and earning a comparable income has to pay 1,620 dollars in income tax and social benefit contributions), and relieving the tax burden of families has to be continued. In 1990 this type of reform of the tax structure will be much more urgently needed than a few, isolated, marginal amendments in the meantime.

On the other hand, the possibilities for stimulating the German money market to take on more venture financing have not been exhausted, by any means, and could be put into practice much more quickly. The establishment of a second stock market with lower issue requirements than at the official trade market is long overdue. Shifting the emphasis from playing it safe to being more venturesome in government supported investment plans would also be a measure that would not adversely affect the consolidation course of the government in any way. At the same time such measures could get people thinking about their own stock and bond investments and loss allotment shares in looking around for means to invest.

We have already repeatedly discussed the fact that the new technologies will bring about basic changes in the sectors of our economy. Not just social and cultural policy but also the public policy of free enterprise has to devote its attention to this phenomenon. If one loses sight of the fact, namely, that the importance of the manufacturing sector is giving way to a predominance of the information orientated service industry, it is almost inevitable that attempts will be made to preserve the old industrial structures and give them unwarranted preferential treatment at the expense of the new industries. The tenacity with which subsidies continue to be granted to branches of industry which are no longer of vital economic interest clearly proves that this danger exists.

If the free enterprise term "industrial policy," with its emotional implications, still appears of relevance, then in this connection. The fact that industrial policy has taken the stand of supporting modern technologies, although their field of application lies more in the realm of production orientated services rather than in industry itself, is indicative of the whole dilemma and theoretical confusion in which we find ourselves. The radical changes in structure have given lie to the work of some economists who have subsribed to the mechanism of a tightly controlled economy.

For the first time in 1978, the Federal Department of Trade and Commerce commissioned a report on the structure of the economy which was undertaken by leading economics institutes with a great deal of time, money and energy expended. The recommendations based on their findings did not differ in any great degree from the annual proposals by the Council of Economic Experts for the development of the economy as a whole. This is probably due to a number of mental reservations which are difficult to overcome: that structure predictions, no matter what, should not be interpreted as an evaluation of growth in the different branches of the economy; that a secular decision between a service society and a purely industrial economy has not been made yet and far-reaching conclusions have to therefore be avoided; and finally that the politicians and the population as a whole are much more interested in the question of the annual growth rate to be expected and therefore economic policy advice takes priority over other considerations.

As I see things, however, the question of the development and redefinition of the term infrastructure is of crucial political importance, without which the process of sectorial change cannot be understood. By the same token, one has to recognize that a modern economy can no longer thrive with just the three traditional production factors; that means that we have to accept the fact that in the future other infrastructures will play the predominant economic role than those of the past or present. Infrastructures are nothing more or less than public services which are a very important factor in almost every branch of the economy. One will have trouble, therefore, in examining the question as to direction and degree of a shift in the structure of the private economy (which cannot be precisely pinned down in any event) without being aware to a certain degree of the changes in infrastructural trends.

By placing too much theoretical emphasis on matters which can only be clarified by actual political practice and experience in life, one ends up very quickly in a blind alley – similar to attempting to determine "public goods." Certain services – such as maintaining and developing the highway system, building schools, hospitals, sport centers, post offices, and telephone networks – have a long tradition of budget support and have been used by many people from time immemorial. As a result, they belong to the category of infrastructure according to popular opinion and, ergo, it is the government's business to take care of them. If the government has always been responsible for regulating something – and the more money and administrative work involved the more evident this seems – it should and must continue to do so. The areas to which this principle of custom does not apply, however, cannot prove themselves to be legitimate infrastructures and should therefore be removed from its sphere of influence.

This kind of thinking creates a structural policy which is orientated on the past and which encourages the government to exercise public control in continuing to produce things which are already available in large quantities and for which the organizational pattern has been determined long in advance. A farsighted structural policy, in contrast, should not be the government's business because in this area the selective process of the market must indicate the direction in which needs will develop.

It is no wonder that this kind of role allocation, in accordance with Parkinson's model, leads to overcapacities and bloated administrations. They serve in turn to confirm the policy that government authorities, under no circumstances , should be assigned new spheres of influence since they have given sufficient proof of their inefficiency in dealing with the areas they are already in charge of. A policy which orientates itself on the past instead of looking forward is self-complacent. It complains about the dynamics of free enterprise being crippled by government structures while at the same time having its share in their theoretical justification, making their transformation in keeping with the times more difficult.

The only way out of this dilemma is to recognize that infrastructures cannot be formally determined in the context of tradition but only materially, in accordance with their actual importance.

To illustrate just how traditional our current concept of infrastructures is, I would like to quote some passages from a leaflet for the purpose of recruiting east German industrial workers into the Ruhr area in the year 1887. I owe my thanks for this document to Werner Maser's book "Am Anfang war der Stein" (In the Beginning was the Stone).

"Masurians!

In the Rhineland region, surrounded by fields, meadows, and woods, the preconditions of good air, lies, just like a Masurian village, remote from the hurried bustle of the Westphalian industrial area, a charming, completely newly built settlement of the Victoria coal-mine near Rauxel. This settlement consists for the time being of over 40 houses and will later be increased to about 65 houses. There are four apartments in every house, two upstairs, two downstairs ... The rent for 1 room (with barn and garden) is just 4 Marks a month, cheap in any event by Westphalian standards. Furthermore, the mining company gives a refund of 1 Mark per month to every boarder ... Beautiful, broad streets run through the whole settlement, water taps and sewerage are available. There is electric street lighting. There is a small garden in front of every house in which flowers or vegetables can be grown. There is a prize for the most attractive garden.

A cooperative store will also be established within the next few years in the settlement where all kinds of grocery goods, such as salt, coffee, herring etc..., will be supplied by the company at reduced prices. A butcher's shop will also be established. For large purchases Castrop, Herne, and Dortmund are nearby ... Schools have also been built for the children so that they do not have to walk far. The miners also do not have more than a ten minute walk to work. The next railway station is about a half hour away. Those who are thrifty can put their money into the savings bank. Many East Prussians have been able to save several thousand Marks in Westphalia.

Masurians, the mining company is above all interested in getting nice families to set up house in this new settlement. The moving costs of every family will be fully reimbursed as well as the travel expenses of single men. The mining company does not want to entice anyone to leave their native country or to leave their loved ones behind, it just wants to give decent people, who cannot find work in their own country or who do not earn enough, the opportunity to make money and to save a little extra so that they do not have to starve in their old age."

Almost everything which is counted as infrastructure today can be found in this appeal: reasonably priced apartments, good streets, convenient shopping facilities, neighbourhood schools, direct railway connections, not to forget a healthy environment. The mining company's management even gave thought to the social integration of the "foreign worker" – possibly earlier and more thoroughly than was done in some places during the reconstruction phase of the Federal Republic of Germany.

When communities advertize industrial settlements nowadays, they basically follow exactly the same pattern.

The question is whether these criteria still correspond to the needs of the companies and their employees. The more a company is involved in the area of high-technology, the greater a role other factors play in deciding where to build new plants than they did in the past: What research institutes are nearby? How far is it to the nearest university and what are its main areas of research? What about the cultural life in the surrounding area? Are there attractive recreational and sport facilities in the region? How long does it take to get to the nearest airport? The main question, though, is whether the company will be able to find enough highly qualified employees.

It was along these very lines that the Japanese government, in 1980, chose fourteen locations, among a number of candidates, for futuristic "technopolis" centers in an effort to break up the concentration of industry in Tokyo and Osaka. Only areas that could offer guarantees for the establishment of an industrial sector involving intensive and technologically orientated research were considered. Universities, public and private research and development laboratories, as well as adequate cultural facilities were supposed to be in the area. Experience had shown that transplanting qualified personnel and their families in areas without sophisticated service infrastructures was difficult. That is why only communities within commuting distance of cities with populations of at least 200,000 were considered. In the past, MITI, the important Ministry of Foreign Trade and Industry for guiding economic development, had given preference to coastal areas in order to guarantee quick access to sea routes for the transportation of raw material and finished products. In contrast the "technopolis" concept completely ignored such criteria. Since the new products were more valuable but smaller in size, one reasoned, being near an airport or highway network was sufficient.

Modern infrastructure is much more than just a collection of certain structural conditions such as railway tracks, streets, sites that have been developed in a spatially limited area. Modern infrastructure is a complicated socio-cultural network which has to take, not only the technological prerequisites for production, but also human value preferences into consideration. The better one succeeds in establishing both together, the higher one can estimate the quality of the location. It is the government's responsibility, which no one surely can still dispute, to provide suitable infrastructures, not only in terms of living conditions, but with regard to economic activities as well. Particularly a government which is interested in making the best use of tax revenues should not be denied the opportunity, therefore, of interpreting infrastructure as a materially dynamic, instead of a formally static, term and making this the basis of its investment decisions.

The volume of revenues, unfortunately, does not automatically increase as new measures are called for. This means that priorities in terms of who most needs support must be continually weighed and reflected in the form of increases or decreases in funds. The establishment of an informational infrastructure should clearly take precedence over other considerations since there is such a large deficit in this area at the present time. At the same time, developments must be controlled by a sensible regional policy so that we can take advantage of existing assets and compensate for discernable weaknesses. It makes sense to begin the establishment of a network of glass fibre cables in densely populated industrial centers, but it is absolutely necessary to rapidly expand the development into rural areas in order to avoid the establishment of a new provinciality of non-information. The possibility of having access to computer centers and data banks is particularly of value in middle-sized cities since it can act as a magnet on young entrepreneurs in special high-tech branches who want to profit from the lower real estate prices and rents in comparison to those in large cities. There are signs that technologically orientated colleges could play an increasingly important role in training management personnel for middle-sized industries, just as the technical trade colleges once did.

The active cultivation of the arts, culture, theaters, galleries, museums, orchestras, and ballet troupes, however, are an important element of this development. It can be more important for a metropolis to improve the quality of its cultural life, for example, than to add another research institute to the ten others that already exist in the area. An infrastructure policy in keeping with the times does not consist of simply supporting new building projects or expanding scientific and technological capacities. It has to be a mirror of the needs which are expressed by people in their individual examination of questions pertaining to the environment, society, and technology. To the extent that these needs can be satisfied, not only economic support, but also cultural and social integration can be achieved.

Those who feel that this line of thinking is too pragmatic, might perhaps find comfort in having their attention drawn to the fact that the necessity of supporting an information and communication technology orientated infrastructure can also easily be justified from the point of view of a "classical" concept of state guidance. The anti-trust commission, for example, in a special report dealing with the role of the German Federal Post Department in telecommunications, emphasized that "the establishment and maintainence of telecommunication networks can be classified as infrastructure and therefore fall within the government's sphere of providence." Since within this framework certain services should also be available on a widespread basis, even if particular economic considerations are an obstacle, "the guarantee of their being supplied on an extensive scale must be ensured by appropriate measures by the authority of the government."

One must also regard an extensive telecommunications network with its connecting links to data centers, administrative institutions, companies, commercial service industries, electronic media, and the new forms of office communication systems as just as important as the traditional links, to the extent that they serve to transmit information. As the service share in the total economic production grows, one will have to acknowledge the electronic communication media as functioning as a means for transporting goods. Furthermore, one cannot deny that the total educational system – with the possible exception of the elementary schools – is affected by information technology.

According to generally agreed upon estimates, about seventy per cent of the working population will have to have a basic knowledge of computers by the year 2000. Even if one adheres strictly to a free enterprise concept of the government's sphere of responsibility, there is no getting around the fact that a basic knowledge of information technology will have to be added to the existing educational infrastructure. Finally, if one considers the enormous potential of the new technologies in the area of environmental protection, the idea of the government not just being responsible for establishing emission limits but also for providing pure air and clean water by means of a "smokeless industrial policy" is certainly not far-fetched.

In discussing the theme of infrastructure, it would be amiss not to at least touch upon the subject of foreign trade in this connection. The increasing inter-connections between the highly industrialized economies is a truism. The volume of foreign trade grew on the average about twice as fast as domestic production in recent years; the international financial markets have been expanding at an annual growth rate of up to thirty per cent since the seventies. This expansion in the exchange of goods and financial transactions is directly connected to quicker and better means of international communication and transportation.

On the other hand, this has given rise to a growing tendency to achieve or maintain international positions of economic leadership by means of information protectionist measures. Since the production factor "knowledge" is an inexhaustibly rich one, but also a very fleeting and rapidly changing good, the attempt is being made to keep it confined within the national borders. In view of the magnitude of economic factors involved, the matter is anything but child's play. When the European Community refused to allow the American information industry access to the European data network Euronet, talk about waging an information war could be heard on the other side of the Atlantic. While international travel is generally becoming easier, the flow of data over international borders is clearly being subjected to more and more restrictions.

"The United States," the *Zeit* wrote in late 1984, "has a unique monopoly on the information market ... Over 95 per cent of the scientific and around 85 per cent of economic information is sold there while the Federal Republic of Germany can only claim a slim 0.6 per cent, respectively 1.4 per cent; although the Germans produce 14.3 per cent of the gross national product of Western Europe and of the USA, they do not have more than a 0.9 per cent share in the corresponding specialized information market ... Should the USA limit the export of data banks in the future by declaring certain areas as "top-secret," as they are already now trying to do with high-tech know-how, Europe, depending as heavily as it does on American data banks, would be put in a very awkward position."

Establishing our own information systems is therefore an indispensable prerequisite if we want to continue to keep up with the international competition. No matter what we set up in the way of national economic structures, their ability to function depends in the final draw on not being damaged in their essential elements by any kind of external "information crisis." Securing supplies, an argument which is often bandied about when it comes to coal subsidies or our allegedly excessive dependency on foreign sources of energy, should in truth cause us a lot more worry concerning the area of information technology. In terms of public policy, in any event, there can be no doubt that a competitive research infrastructure is the means for guaranteeing both our economic and social well-being.

The question as to what extent the market is "liable for values," which we named as the third problem in connection with the changes taking place in economic and technological structures, was made a matter of public awareness with the discussion of the "Limits to Growth" in the early seventies. The controversy was quantitative versus qualitative growth, the outcome, so one supposed, providing an answer to the social and political sixty-four-thousand-dollar question: "What do you think about free enterprise?"

There is good reason why nowadays, barely ten years later, this debate no longer plays a very important role, with the exception of a few consciously polemic attacks. It soon became apparent, namely, that in practical terms the controversy, which was fought with dogmatic enthusiasm, is meaningless. What the one side denounced as quantitative growth was just as much or as little a matter of quality as their own ideas about what the "right" growth aims were; they simply had different value preferences. This, and the fact that they were in the minority at first, created a certain elitist attitude of arrogance towards the allegedly thoughtless consumer orientated thinking on the part of the majority. On the other hand, an unnecessary market situation was fabricated which would have had to be completely value-free in order to be able to function and, as a result, contained dubious amoral elements. In reality the market is not value-free but a very reliable mirror of society's and the individual's needs and wishes; and these can change so much that a position held by a minority can suddenly become a stance of the majority.

Both sides have obviously learned a lot in the meantime so that the conflict has lost much of its earlier acrimony. Those who flirted with the idea of zero growth have had to recognize the fact that without some nominal increases there cannot be any qualitative growth in their sense of the word. Those who wanted established goods and services, as reflected in the gross national product as a general index of economic activity, to be the only valid criteria of growth, have had to acknowledge that actual changes in values and structures cannot be pinpointed in this way.

The whole discussion could simply be dropped if it were not for the fact that the gross national product still plays a role as an indication of social well-being and as the basis for political decision-making, although this function is out of all proportion to its actual significance. It has often been pointed out that many things (for example, the work of housewives, artistic/musical works, private caretaking) are not included in this allegedly comprehensive account or are incorrectly audited (accidents and environmental damages as economic growth factors). The tendency to overrate the old G. N. P. criteria is gradually becoming a policy problem as more spare time is spent in economic or socially orientated activities, as more and more people are taking advantage of adult education opportunities, and as we are increasingly successful in conserving energy and protecting the environment. The annual per cent point predictions, which are announced with jubilation only to be often revised later, do not tell us anything about either the increase in scientific knowledge, the use of technology, the coping with social problems or environmental treatment. In short, they do not say anthing about the actual state of affairs and are not reliable indications for political action.

In a year of high national product growth crucial mistakes in structural policy can be made while getting things on the right track towards a long-term safeguarding of the future can be done in a year with a low rate of increase. The signals being sent by this out-dated index of the state of the economy can prove to be more of a hindrance than a help in shaping a long-term policy for structural changes.

The employment market controversy, which flared up in all fierceness in 1985, clearly indicated that people are no longer willing to attach as much importance and political meaning to global figures as in the past. As long as there is mass unemployment, many citizens do not see even above average growth rates as an indication that things are taking a turn for the better. This is the revenge for having had the structural determination factor take second place to economic theories when explaning unemployment in the political debate for many years.

Especially in view of the strong public pressure on the government to prove the success of its economic policy by creating more jobs, it is evident that a purely free enterprise model, as we have already proved on hand from a number of historical examples, is unrealistic. The government for many years now has taken upon itself the obligation – and for decades the business community, unions, and the economists have encouraged this policy – to try to compensate for the negative effects on the employment market resulting from the free enterprise power play between the supply of available jobs and the demand for work. This was attempted, first of all, by artificially stimulating the demand side, whether in the form of employment programs or by increasing the number of government employees. After both methods have proved to be unfeasible in the long run, boosting the private supply side is now generally being called for. If specific infrastructural support measures, which small and middle-sized companies and new service industries employing a proportionately high number of people, in particular, will be able to take advantage of, are not part of the plan, however, it is obvious that we will again risk wasting our limited public revenues.

What we are basically dealing with here is the need to gradually develop a policy of employment which is no longer completely dependent on an outdated pattern of bipolar systems of order – the economy on the one side, the goverment on the other –, but one which is integrated into a smaller, decentralized, and flexible system of organization. There is no such thing, for example, as *the* employment market. In the mid-Neckar region there is almost full employment and even a serious lack of skilled workers for the regional mechanical engineering and automobile industries, while barely a hundred kilometers to the northwest, in the Odenwald, lies a region which can be classified as structurally weak.

An infrastructure policy which is technology orientated and geared toward middle-sized businesses might be even more effective in influencing the employment market than general control measures. Furthermore, it has been clearly shown that more jobs have been lost in the last decade by the failure to adopt modernization measures rather than by their implementation (in a 1983 study by the German Institute for Economic Research it was established that 82 per cent of the total job loss occured in branches with below average productivity standards). A preventive employment market policy boils down therefore to an infrastructure policy of supporting structural change. Those who do not want to concede this to the government can hardly expect the state, on the other hand, to take on any more responsibility in the field of employment policy. Simply being responsible for removing the damages done by economic processes is demanding too much (as the volume of subsidies shows) as well as too little (in the sense of just doing patchup work) of the government's potential. One may point out in this connection that almost half (40,700 of a total of 90,900) of the new jobs in the Federal Republic of Germany in 1984 were created in Baden-Württemberg, where the structural change is being pushed forward by an active economic and technological policy, as is well known. The number of new jobs in the service industry was also about double that of those in the manufacturing sector.

The controversy concerning the question of the quality of growth has died down; this does not mean, however, that it will no longer play a public policy role in the future. It is necessary to be prepared in time for developments that will arise in connection with technological progress. Let me briefly clarify this point quoting three examples:

First of all: The generally higher demands for environmental protection will inevitably be reflected in the prices for goods and, therefore, to a certain degree, in the way private income is spent. If we do not want our lakes and rivers to be polluted by industrial wastes, there is no way of getting around the fact that we will have to pay for industrial recycling processes in higher prices for things. If we want to ensure our supply of energy and still have clean air, we will have to pay for the desulphurization and denitrogenization facilities in the form of higher electricity bills. If you do not want to give up your car and yet are concerned about saving our forests, you will have to dig in your pockets and come up with the money for a catalyst.

There will be similar cost transfers in many other areas. They undoubtedly will considerably narrow down the general consumer buying potential. This will not be the result of any kind of government control of growth in selecting quality features, however, but simply the outcome of a change in value preferences in line with market requirements, namely, in the form of higher prices for better quality. It is the job of those in power to make this perfectly clear, with all due emphasis and in good time, in order to avoid an eventual new escalation of demands according to the motto: Once the environment has been cleaned up we want to be able to afford our old status symbols again. Now that we are determined and have come to realize that we can no longer afford to stick to a collision course with the environment, we must be willing to adapt our consumer behaviour accordingly.

Secondly: As the application of gene and biotechnological research findings becomes more concrete and varied, the ethical questions concerning their commercial exploitation will become more and more pressing. The range of biotechnical procedures and products, which will be flooding the market by the year 2000 and thereafter, can scarcely be estimated today. Among them will be many products which can certainly be welcomed without reservation: highly specialized biological pesticides; biotechnically manufactured nitrogen fertilizers; useful plants that have been made more productive and resistent by means of gene manipulation; bacteriological "recycling factories;" microbes for obtaining industrial protein and sugar; more effective and cheaper medicines; new possibilities for fighting genetic deficiencies.

But a grey market of frozen human embryos could also emerge, and concerning their constitutional status there is a great deal of uncertainty under the current laws in force since nidation (nesting) has not yet taken place. Animal breeds could be offered for sale which would be difficult to bring into accord with our understanding of creatures made by God. Patents have already been registered in the USA for gene surgery procedures which are supposed to be able to increase the growth of animals tenfold. Private clinics could flourish which would do a brisk business with the hopes of the fatally ill by risky manipulations in the genetic make-up of the patients.

It is not enough to delegate all the responsibility for these problems to the scientists themselves. Even if we can assume that a functioning scientific code of ethics rules out the possibility of unwarranted manipulation of human genes, there are still a number of conceivable procedures whose commercial exploitation would run contrary to our current understanding of nature. The question of just who is to profit from the biotechnical progress also cannot be dismissed as value-free by alluding to supply and demand. Is it conceivable that the industrial nations, who already do not know what to do with their food surpluses, will continue to increase their yields by means of agricultural "test-tube crops" while the developing countries, with their out-dated farming methods, sink further into hunger and misery?

The headway being made by ethical and normative postulates in shaping our economic and social co-existence (or, as one could also put it, the inexorable dismantling of the pure Homo Oeconomicus as the ideal of social policy), which can already be seen in the area of micro-electronic information technology, will really take off the ground within a few decades when our industries are organized along both micro-processing *and* biotechnical lines. A society which has not learned to also include ethical rules of behaviour for individuals in its economic life, could, despite a high standard of material prosperity, end up in an intellectual and cultural ghetto.

Thirdly: Recently the discussion about values has been increasingly seen in terms of the civil or military uses of the new technologies. The discussion apparently generally centers on the question of research support and not directly on the market as such. The matter can be seen as directly relevant to economic activities, however, to the extent that research capacities, in view of their short lifespan nowadays, also act as production stimulants. This is even more apparent in the foreign trade sector: The embargo list of Cocom (Coordinating Committee for East-West Trade Policy), whose members include almost all the NATO countries as well as Japan and whose job it is to prevent security relevant products from being exported to the nations of the East Block, now has more than 150 items listed, almost all of them high-tech products.

Technological progress nowadays is generally ambivalent – it can serve to improve medical equipment as well to perfect weapon systems. Exercising a kind of research censorship for this reason, as some people seemed inclined to do when it came to discussing Europe's participation in the American SDI research programm, is not the solution to the problem. Just as in the social area, we would be placing a burden on the future generation – without asking whether they were in agreement or not –, expecting them to cope with "becoming a third class industrial country, left sitting in the squalor of out-dated industries" (Peter Glotz). Ethical conflicts are seldom resolved by sacrificing the acquisition of knowledge; more often they demand a politically and socially responsible control of the uses of knowledge.

The railway was developed so that people could get to places faster, and it was an essential infrastructural support in establishing the modern welfare state. The first tracks had barely been laid, however, when it was being used in Prussia as an instrument for mobilizing troops and in Austria for preparing an armed encounter. The mistakes and weaknesses are not in the quality of our technological discoveries but in the manner with which they are dealt with; in other words, we have only ourselves to blame.

The question of growth, which one used to believe could be handed over to the market, the environment, the government or some other kind of institution, is being ruthlessly brought directly home to us by the new technologies. In technically reducing the dimensions of time and space, they inexorably force us to face the question of the dimensions, in terms of time and space, of our own moral consciousness.

Chapter 17
More than Just a Vision?

A Society of Reconciliation

To start with we have to again stress the fact that our large-scale system blocks, which have emerged over a long period of time and have depersonalized our sense of duties and responsiblities, are on the verge of collapse. Not society as a whole, but the ability of its large-scale systems to solve our problems has reached its "Limitations to Growth." The new technologies are capable of creating decentralized structures with more self-reliance, a wider sphere of responsibility, and improved environmental protection. For this opportunity to become reality, though, we have to be willing and able to make a break. This calls for a policy which no longer wastes its time solving antiquated conflicts with out-dated patterns of order.

What are our chances of steering change in this direction? The existence of polarized power structures is, at the start, more likely to provide new cause for conflict rather than to ease rapprochement. Once institutions have been established that involve employing a large number of people and a great deal of expense, they tend to lead an independent existence which can be pretty far removed from the realities of society. Even if the original unsatisfactory state of affairs has long been rectified, the factions involved in the administration of such institutions violently resist any efforts at changing the system or the allocation of public benefits. There could be any number of unoccupied private apartments and some people would still be demanding that new public housing be built; there could be half the number of school aged children and one would still read reports about the alleged shortage of teachers when it became time for fall hiring. Established patterns of conflict, which enough people have made it their business to worry about, show a tendency to become a "never ending story."

Without political and social support even such a vitalizing, dynamic force like technology has difficulty in replacing out-dated structures with new ones. Innovations are simply possibilities, opportunities which one can take advantage of or not, as the case may be. That is to say: Without the necessary will to change things culturally, without a basic "intellectual trend" towards renewal, even the most positive technological options will come to nothing, even the most sensible attempts at political reform will end up in failure.

Remarkably enough, however, there have been a number of recent developments which give cause for hope. Indications that fundamental changes are underway can be seen in a variety of different disciplines and walks of life, a willingness to overcome old conflicts and to see things from a completely new perspective. With all due caution and reserve which one must exercise in judging contemporary movements of which oneself is a part, I think one can permit me to remark that the fascination with a society in conflict, which sees its only source of power in analytical criticism, is over and done with. Social conflicts are neither being cultivated (as they were by the student generation of the late sixties) nor irrationally demonized; the idea of conflict, allegedly being an inevitable aspect of history or fate, is giving way to a theoretical and practical search for solutions. The concept of integrity, and within this sphere the reconciliation of conflicts, plays a very important role here. In the following I would like to name some factors which I see as symptoms for this trend while expressly reminding the readers of what I said at the start about the "necessary dilettantism" of politicians in areas that do not belong to their field of speciality.

Among the natural sciences, the claim to a position of philosophical leadership by the physicists has become more and more audible. The science which, like no other, has been so instrumental in taking apart, even in destroying our conception of the world, obviously sees itself as having arrived at epistemological limits which call for a fundamentally new philosophical consciousness. To what extent this is still the conviction of a minority or is already the majority's opinion, I am not in a position to judge. In any event, there is a visible body of opinion calling for a philosophy of the macro-processes on earth which does not completely bypass the new micro-cosmic discoveries.

Apparently quantum physics play an especially important role here. It is not possible to determine with exact precision the position and time of an elementary particle (Heisenberg's indeterminancy principle). In the subatomic region our concepts regarding matter and non-matter fail to function as well as our usual way of thinking about causal relationships. There is not any fixed pattern of things or movements in space and time but only "probability structures," non-local factors of influence whose appearance can neither be explained or predicted with our present standard of knowledge.

Modern physics, on the basis of Einstein's theory of relativity, shook the apparently unshakable basic human parameters of space, time, matter, and causality to their foundations. In recent years, if I read things right, even the fundamental principles of science have been getting a pretty rough raking over the coals. The fact that in the world of the smallest particles the arrangement of the measuring apparatus is capable of influencing the readings (without there being an objective mistake in the sense of errors in measurement), has lead physicists to doubt the existence of unchangeable "truths" in the laws of nature. Thus mathematical formulas are only probability demonstrations but not fundamental facts of nature which are completely independent of man.

This in turn has led one to conclude that the philosophical view of a world which can be analyzed as much as one likes and which functions according to the principles of deterministic rationality – in the way that was established by classical physics – is no longer tenable. On the contrary, one has to think in terms of "integral" categories of continual dynamics of correlations and interrelationships in which man is necessarily an integral element and from which he cannot simply remove himself to act as a neutral observer.

I have taken up this theme in the awareness, as a non-expert, that my presentation of developments could comprise misunderstandings or misinterpretations. The fact seems very significant to me, however, that in a field which is generally regarded as one of the most highly specialized, detailed and logical of the sciences, there is an impulse to see things from an interdisciplinary point of view and with reference to philosophy and ethics. Without going into raptures right away about a new age, one has to admit that the accusations concerning the fragmentation and lack of self-reflection in the modern sciences certainly no longer apply to physics. This branch of the natural sciences is in fact now one of the strongest supporters of an ethics of responsibility which is orientated to humane forms of criteria and in gear with a reconciliation between the economy and the ecology.

The humanities tend to react with mistrust rather than sympathy to this blurring of boundary lines between questions of knowledge and orientation by the physicists. The separate lines of development which the natural sciences and humanities began to take two hundred years ago have led to great differences not only in their subject material but also in their scientific methods of research. As Carl Friedrich von Weizsäcker observes with regret in his 1964 book "Die Tragweite der Wissenschaft" (The Scope of Science): "The humanities and natural sciences scarcely have a language in common in which they could even talk to one another, and, as often as not, they are both even proud of this alienation." The natural scientists seem firmly fixed to a course of examining matter in keeping with Galileo's law of science "measure everything which is measurable and make everything measurable which is not yet measurable," while the humanists claim for themselves the exclusive authority to explain man's existence from a cultural and social philosophical point of view. A change is becoming apparent here too, though, a return to the "Einheit der Evolution von Natur und Kultur" (Unity in the Evolution of Nature and Culture) [Werner Conze]).

For the first time in the fall of 1984 large numbers of natural scientists participated in the 35th Historians' Congress in Berlin. Talks on parallel developments in biology and culture as well as introductory lectures on theories of physical structure were part of the congress programm. The annual session of the West German Rectors'Conference in May of 1985 was devoted to the theme "Demands and Challenges of the Humanities." The conference reached the conclusion that everything had to be done to overcome the division between the natural sciences and the humanities and to support interdisciplinary projects and dialogue. This change of attitude was obviously fostered by the natural sciences confessing that only "partial patterns of order and partial truths" (Werner Heisenberg) could be discovered in quantization processes in nature. In addition, the humanties now also have a research method at their disposal, cybernetics, which has proved to be valuable in the area of information technology as well as in social and natural science fields (particularly in explaining natural, recurring cycles).

Once the humanities become more realistic in their approach, and there are indications that things are taking a turn in this direction, and no longer ignore the economic and technological realites of life, they will be able to regain their position as a cultural force of integration. Within the near future, of course, the natural sciences and technology will continue to be more visible than the humanities in influencing the "vita activa," the active life. Even in the face of this fact, however, the humanities cannot simply retreat to their ivory towers of academic life and give themselves completely up to the "vita contemplativa."

As a steady, critical partner of technological progress, appraising it in terms of its chances of achieving worthwile values for the individual and society as a whole, the humanities could become a very important factor in stabilizing our culture. For the growing flood of technological information they could establish cultural and anthropological standards for humane information processing. They could change communication capabilities into possibilities for understanding one another. It is practically a disgrace, for example, that in almost every highly industrialized country (with the exception of Japan) the inclination on the part of young academics to occupy themselves with foreign languages and culture is rapidly disappearing as the amount of available information and data grows. This clearly indicates that information networks alone do not necessarily mean progress in quality. They are like streets in which one can meet other people or also avoid them.

Whether dialogue or monologue, contact or isolation, consensus or dissent predominate in a society, is a question that is decided in the preliminary stages of every technological manifestation. The humanities are by no means responsible for repairing the damages that technology does to our society and culture. Aside from the fact that they could not even do justice to such a commission in most cases because the "normative force of reality" would have already carried things much too far, they would also be failing to take advantage of what is their own true cultural potential. Nothing is more open to the future in this world than the human intellect. When talking about technological options one has to continually keep in mind that these options are actually limited and restricted, selections from an apriori spectrum of creative intellect. To make a start here by putting things into perspective and helping to shape developments by also considering the new technologies in terms of their function as instrumental and infrastructural reinforcers, would be infinitely more helpful than persisting in the historical confrontation between disciplines which have been alienated by a great deal of polemics. Now that the natural sciences are focusing more of their attention at things from a comprehensive point of view, perhaps not in terms of their methods, but certainly as far as the finality of their interpretations go, the humanities should be all the more willing to venture a new start in the direction of interdisciplinary cooperation.

The discussion about questions of ethics, which I am pleased to say seems to have been undergoing a new revival in recent years, could provide important stimulations and impulses. Our experiences with the vulnerability of our environment and the global long-distance effect of excessive emissions have given rise to the urgent call for far-reaching ethics of prevention and care. As we have come to recognize the way in which ecological systems act in cyclical combination with one another, we have become more aware of our own interdependency on the natural environment. A new sense of responsibility is the logical consequence of this awareness and two tendencies are indicated: on the one hand, a more personalized sense of one's duties and failure to meet these duties and, on the other hand, the global scope of these responsibilities.

One must, though, take care not to place too much of a strain on people's ideals. One certainly cannot expect a sense of consideration beyond the point of people's own legitimate interests. Turning our backs on progress and establishing a society which is not fit for humans is not the purpose of either creation or culture. The demand for an integral "ethics of responsibility for the future" (Hans Jonas) is nevertheless justified. It is also of pragmatic value in that it makes the deficiencies of an individual enforcement of ethics especially apparent. How can we hope to credibly establish a broad sense of ethics regarding nature and the environment if we still have trouble in dealing with personal virtues such as charity, brotherhood and consideration for others? (This is also the main dilemma of the Greens in establishing a sense of credibility. They are so intolerant in their dealings with people who think differently from them that the discrepancy between their postulates for a sense of environmental ethics and the reality of their sense of human ethics is often frightening.)

There clearly are parallels between this development and a new understanding of human freedom which we already talked about in other contexts as being a fundamental characteristic of the development of information technology. The will to develop a sense of global ethics presupposes a concrete sense of individual ethics. A sense of integral responsibility begins with limiting individual freedoms on a voluntary basis. As the industry begins to switch more and more to environmentally friendly forms of production – and they are energetically in the process of doing so – the more a sense of responsibility on the part of the individual for his/her environment will come into focus.

Once industrial fumes have been denitrogenized and desulphurized, it will generally become very apparent just what part private oil and coal heating play in polluting the air. By the time large-scale technological recycling processes and special treatments have satisfactorily solved the problem of industrial waste products, the crisis in overburdened communal garbage depots, which will not be able to keep pace with the growing mountains of household refuse, will probably just be reaching its climax.

In the future we will be facing the challenge of coping not so much with the dangers of technology but rather with its success, a success which will be very expensive and which will force us to examine the earnestness of our principles as it directly affects our everyday life; whether we are really willing to spend more money for environmental technology and going without certain things we would like to buy; whether we are really willing to accept a minimum standard of ethical behaviour which is of service to the community in those areas where technology does not have a positive effect on reducing damages and the government is not responsible for establishing a standard of norms, in the highly personal sphere of the way we conduct our lives (this applies equally to health factors – one only has to think of the damages caused by unhealthy consumer habits which run into billions of dollars).

It is not a matter of chance that the categorical imperative plays such a central role in the modern discussion of ethics. "Act in a way that the consequences of your actions are not destructive for the continuation of human life on earth as we know it" is the frequently quoted precept from "Das

Prinzip Verantwortung" (The Principle of Responsibility) which was published by the philosopher Hans Jonas in 1979. In the same spirit is Günter Altner's axiom "Act in a way that the life interests of our current and future ecological systems can be guaranteed."

By rights, of course, one can only demand that an individual acts in accordance with an expanded sense of ethical responsibility in terms of space and time if he or she is not regarded as being isolated from society, as an outsider or even trouble-maker. A government which is one-sided in the individualization of rights and the collectivization of duties, which is actively involved in trying to shift the blame for things, which wants to try to completely subsidize necessary adjustments away – this type of government is sending exactly the wrong kind of ethical signals. It reduces a personal sense of ethical responsibility to a non-entity and is then surprised at the explosion of costs resulting from the egocentricity, the lack of social and environmental consciousness on the part of its citizens; the citizens in turn complain about the increases in taxes and social contributions. It turns out that an individual and a global sense of ethics are inseparably bound together, in a negative as well as in a positive sense. If personal virtues were to receive the support they deserve by government patterns of order in their function as social models, people would voluntarily extend their personal sphere of responsibility. If policy, on the contrary, works to counteract against criteria belonging to an individual sense of ethics, not only does the willingness to assume community responsibilty disappear, but this lack of responsibility eventually has a negative effect on direct relationships (family, neighbours) and can even lead, in extreme cases, to a reversal of values, to a "negative sense of ethics."

The ecology movement, when it was not wasting its time in superficial actionism, was able to recognize and point out this interrelationship with gratifying clarity, couching its postulates in such terms as "an integrated point of view" and "a global sense of ethical responsibility." However, as long as there is no echo in society and culture on the whole, an ethics of responsibility for the future will be left holding the baby on its own. This is because people are not only simultaneously in contact with both their natural *and* social environment but also guided by the same moral principles in dealing with both.

The fact that strategies of social conflict are no longer as appealing is an indication that the time is ripe for an "integrated social ethics" which has the support of both politics and science and which would be able to draw strength in establishing a new frame of order from the morality of the individual. One cannot help overlooking that we are currently undergoing a shift in values while at the same time new technological options are presenting themselves, and the government is showing signs of no longer being able to cope with the burden of its program of welfare benefits. Only a new social policy can rectify matters which makes the establishment of a new frame of order not only plausible, but a matter of urgency.

Is it possible that all these reflections on the position, duty, and responsibility of man in modern times can pass politics by without leaving a trace? Are party traditions and habits of conflict so deeply rooted that they can remain untouched by the new currents in philosophy, culture and ethics? By just looking at the daily headlines in the media, this would seem to be the case. But politics is also aware of the evolutionary change, you just do not read about it on the front page of the newspaper.

In particular the two large democratic parties, the CDU (Christian Democratic Union) and the SPD (Social Democratic Party), have undertaken marked course corrections with regard to domestic policy in comparison to the positions they held just a decade ago. After some initial resistance the Social Democrats are gradually becoming more and more receptive in regard to the new technologies. "Technological innovations make new products, new production processes and new services possible" writes, for example Hesse's minister for trade and commerce, Dr. Ulrich Steger (SPD), in an article in the *FAZ (Frankfurter Allgemeine Zeitung)* supplement, *Blick durch die Wissenschaft (A Look at Science):* "They can ensure the Federal Republic's maintaining its position in international competition and thus contribute towards creating secure jobs. The ecology is a prime example of the way in which technology can at the same time be a factor in contributing to our problems and the key to solving them; technological innovations offer the opportunity of bringing about a reconciliation between ecology and the economy. Technological innovations have at their disposal the potential for humanization by which means work conditions can be improved, the status of jobs enhanced and more latitude created for reducing working hours while at the same time guaranteeing incomes." There is no one in the Union party who would not be able to sign this statement without reservations.

Christian Democratic politicians, on the other hand, who were initially very sceptical regarding such terms as "quality of life" and "qualitative growth" have now generally come to recognize the ecological limits of economic growth. Nowadays companies cannot count on being subject to less stringent environmental protection measures or administrative controls in states where the Christian Democrats are in power than they can expect at the hands of the Social Democrats. That the economy and ecology can and must be reconciled with one another is part of the political credo of both democratic parties; and when one stops to think about the original differences in opinion, this is saying a lot.

The alleged conflict between technology and the work world, on the other hand, has not yet been resolved by political debate. The Social Democrats have traditionally leaned towards collective solutions in resolving conflicts so that they naturally have difficulty in perceiving more flexible and individualized employee-employer relationships, of all things, as the most effective means for reconciling the conflict between technological innovation and human labor. Their history is closely tied up with the struggles and triumphs of a tightly organized workforce. Leaving it up to people, even in a limited degree, to make their own employment agreements with a company or even worse, signing individual work contracts, seems to them a betrayal of their historic achievement in having "tamed capitalism."

Yet, in the long-run, sticking to collective patterns for resolving conflicts which are a carryover from the past will not be of any more service to man, than procrastinating in making changes in our industrial structures will help the environment. Making people into inflexible labor suppliers, in contrast to a variable and flexible technology, means robbing them of the most important element of their superiority. Hyperbolically stated, people will end up functioning more and more like the assembly lines of the past: The supply of full-time labor, with all its collective agreement and social restrictions, is the inflexible factor of production while robots and computers represent the elements of flexibility. Businesses will react to this supply in the same way they do to any type of long-term and cost-intensive investment – they will consider whether they can really tie up their capital over such a long period of time and whether the production goals they are striving for cannot be achieved with short-term financial commitments. If they reach the conclusion that this is indeed possible, they will invest in machines – the "shackled" employee ends up losing the race against automation.

Making people more attractive for the production process, by granting them more autonomy in deciding when and under what conditions they are willing to work, does not have to and should not imply a loss of social status. The punch-clock is not a social benefit but an invention of the 19th century for the purpose of enforcing more work discipline. The legal as well as the collective bargaining standards which were afterwards established orientated themselves on this industrial punch-clock rhythm of a collective labor concept because there was no other alternative. Nowadays, in contrast, we have a wide range of technological possibilities and organization models at our disposal for creating a better, that is, more flexible labor structure. They could be instrumental in bringing people's wishes concerning the organization of their work life, on the one hand, and production and service needs, on the other, more in line with each other. The standards continue in existence, however, as if nothing had changed.

There are a number of inconsistencies, for instance, in the regulations for part-time employment: Part-time employment spaced over longer periods of time results in the loss of health insurance and unemployment benefits; part-time work considerably reduces the amount of social security payments while periods of unemployment, which are counted towards pension, do not affect the right to a pension; if a person works less than twenty hours a week his contribution to the gross national product is considered insignificant and he is not entitled to unemployment benefits.

As long as these regulations are in force, part-time employees are in fact socially discriminated. Those who nevertheless enter such employment are "selling" their labor at bargain prices in the eyes of the unions and are, therefore, not acting in solidarity with full-time employees. As a result, the unions do not include part-time employment in their wage negotiations, which the employers naturally do not object to. Many part-time job contracts therefore do not include any supplements, special bonuses, or company social benefits and so the vicious circle is complete. This is how we treat over three million part-time employees and hundreds of thousands of others who would welcome the opportunity of being able to work more flexibly if they were allowed to.

The reconciliation of technology and the work world can only be achieved if both political camps are able to radically change their point of perspective. The Social Democrats will have to accept the fact that within the near future there cannot be full employment for indefinite periods in keeping with the established pattern of full-time jobs with fixed working times. The Christian Democrats also will have to accept the fact that more flexible labor arrangements without suitable proctection against wrongful dismissal and social insurance cannot become the reality of the work world in which the unions also play an important role. Those who want and are calling for making adaptations to the new technologies are on the right track; those, on the other hand, who think to take advantage of circumstances by indiscriminately cutting back on social benefit programs (as has been proposed in some academic circles and recently brought up for political discussion by the FDP [Free Democratic Party]) are blocking necessary changes from being made.

The industrial culture of the nineties will be working with extremely short-lived and highly sophisticated production systems. People with a lackadaisical attitude toward their jobs, working at unskilled laborer wages for want of something to do, are not called for. What we will rather need are qualified employees, who are alert, flexible, and mobile enough to recognize employment opportunities which lie beyond what has been taken over by technology.

Choosing the right control instruments for employment policy is therefore very important because, on the basis of available data, the technological components of unemployment have barely been taken into account till now. Increases in productivity are below the average of past decades while the automation reserves in most branches have scarcely been tapped. This can and will change as modern processing methods begin to penetrate, not just the traditional mechanical engineering field, but also other areas such as wood-processing and the textile industry. In view of the current high unemployment, it is perhaps tempting to take some pressure off the employment market for the time being by creating jobs for the unemployed at "bargain-basement prices." In the long-term, however, looking ahead to the next decade, such a policy would result in a number of fatal consequences.

For one, many businesses would fail to make the necessary investments in technology which they need in order to maintain a competitive position on the international market. Furthermore, the transition to an industrial structure based on capital production, in which the question of wage costs plays second fiddle to increases in productivity which can be achieved by machines, would be considerably slowed down. There would not be the possibility of using automation profits to create new jobs in complementary service functions (maintainence, control, planning, marketing, service and foreign representation) or to finance cultural and social services by means of wages, taxes, and higher revenues. Finally, many employees would be irresponsibly left in the illusion that they do not have to make the necessary effort to adjust to the new technologies, in any way and at any time, as long as they are willing to accept a certain amount of wage cut-backs at their old jobs.

Dealing with the structural employment market crisis in this superficial manner can only be urgently warned against. Developing an employment policy program of do's and don'ts on the basis of Nuremberg's monthly statistic reports (Nuremberg is the seat of the German Federal Labor Office. trans. note) makes just as much sense as designing economic policy according to increases in the gross national product. What is called for is the careful observance of sectoral and regional structures and making this the basis for promoting the development of the economy and technology. The overall aim, "reducing unemployment," must, so to speak, be broken down into operational components: reducing the regional discrepancies between the different employment markets; lowering the proportion of the long-term unemployed; creating new jobs. The former requires, as we have already pointed out, above all a technologically orientated infrastructure. Mobilizing the employment market, in the sense of having fewer people out of work for longer than a year, presupposes the rapid technological development of structural change on the part of the business community; the productivity profits have to be large enough to enable the establishment of new branches of production and the investment in facilities and jobs which are part of such a development. Employees, for their part, must be willing to take as much advantage of government and private educational opportunities as possible, to accept temporary or part-time work as interim solutions, and to possibly try out forms of part-time self-employment in the service sector. It is the government's responsibility to see that the legal and bureaucratic red tape involved in such undertakings is reduced as rapidly as possible. The foundations for creating new, future orientated jobs, as we have likewise already set forth, can be laid by promoting research, supporting technology transfer, and providing aid for new businesses.

I am completely convinced that a consensus of opinion can be achieved between the large political parties, business interest groups and the unions regarding a strategy of this kind for the sectoral and regional mobilization of the employment market, along with a balanced policy of supporting both employers and employees. The new technologies are putting politicians and organizations under a lot of pressure to be more efficient, to allow companies more legal scope and a freer hand in organizing working times and assessing job content. Against the background of the enormous investment costs which are required, the discussion concerning employees having a greater share in the means of production is being revived once again. Everything points in the direction of employers and employees having more freedom in negotiating customized job contracts. The unions would have the job of making sure that certain minimums in allocations and socio-political essentials were included in the agenda of collective agreements concerning conditions of employment. Politics would be responsible for seeing that more flexibility would not mean less social security. In the long run this means, in any event, dissociating the only means for securing one's livelihood from having a full-time job. Dependent employment relationships alone, in the long-term, will not be able to either form the main support for our social security system or to contribute to the total tax revenues at the scale that they now do. Steps in this direction will have to be taken in this decade. Industrial laws and regulations have to be made more flexible and a sense of personal social responsibility must be given new life. This must be followed by a fundamental tax reform and, in the mid-nineties, marked by a general social reform which is based on the new job, social and income structures.

If one keeps the development of the economy-ecology theme in mind, the following prediction is not far-fetched: The new technologies will exert a great deal of pragmatic pressure on politics and society to make use of the advantages they offer to overcome old patterns of conflict, particularly in relation to technology and the work world. The volume of work which will be required at different times will fluctuate to such a degree between the economic sectors and the various life cycles that it will not be able to be distributed on a monopolistic basis (by a limited number of people having full-time jobs), nor on a collective basis (in the form of a wholsale reduction in working time), but only on an individual basis. The unions will find themselves being faced with completely new, as of yet barely tapped areas of responsibility in the form of individual consultation. The political parties will have to learn that old social and economic dogmas are equally incapable of meeting the demand for new regulations. In reading the works of the French communist André Gorz dealing with the liberation of man from the stereotypes of the old worker society by technology, one gets an idea of the potential power of modern information and communication technology to demolish ideology.

The reconciliation of conflicts, which were held to be insurmountable yesterday, is slowly but surely being transformed from a political utopia into a political necessity. Three basic lines of development are becoming especially evident in this transformation of society:

- The loss of stability in collective blocks of power, which emerged in the course of the process of industrialization and which are now faced with the alternative of either completely losing their function or participating in the "transformation by rapprochement."
- The potential of the new technology to individualize life and work cycles, coinciding with a visible, autonomy orientated change in values.
- The structural congruity between an integrated information technology infrastructure and scientific and ethical trends which postulate a new concept of responsibility by recognizing integral correlationships.

Talk about a "society of reconciliation" is therefore more than simply an expression of the subliminal or obvious need for harmony in the face of a world that is getting more complicated all the time and of the exhausting political bickering and quarelling day in and day out. Putting the idea of reconciliation in the forefront rather implies searching for integration formulas that can be moulded into fundamental principles for dealing with social conflicts. The "worst possible disaster", which is theoretically inherent in almost every phenomenon of civilization, should not be the starting point for discussing social policies; this leads straight to a paralyzing immobility, to a powerless entanglement in the contradictions and short-comings of reality, to a despairing spiral of fear and inactivity.

If one thinks, on the contrary, in terms of the "best possible benefit" the political culture acquires elements of dynamic creativity which are capable of overcoming the conflicts and smoothing over strained relations. Raising critical objections has an extremely important constructive function in this process: They are helpful in recognizing and avoiding mistakes in the preliminary stages of the decision-making process so that there is constant pressure to optimize developments. Those who argue on the basis of the "worst possible disaster" do not need to know anything, since they are satisfied with the hypothesis and simply believe in things. This is clearly a case of giving way to ideology. Those who want to achieve the "best possible benefit," on the other hand, have to know and weigh matters since the validity of their thesis can be empirically tested once the decisions are made.

Furthermore, worst case scenarios always lead to polarization because they insinuate the amorality of those who do not go along with them in consciously running the risk of an apocalyptic catastrophe. Best case predictions, in contrast, appeal to both our sense of reason and emotion in achieving goals which can be of service to the community, they demonstrate, in other words, confidence in the moral integrity of our critical fellowmen as well. Conflict strategies necessarily result in either victory or defeat, dominance or subjugation. They do not contribute towards a better understanding between people because the losers are concerned with devising methods to revise the results and the winners will try to prevent this by means of repression. The concept of reconciliation, in contrast, recognizes compromise as the optimal formula for success in that it wants to persuade rather than overcome.

The co-founders of the social market economy, above all Ludwig Erhard and Alfred Müller-Armack, must have been thinking along these lines as they started talking, barely fifteen years after the monetary reform, about the necessity of a "second phase" of the social market economy. Müller-Armack declared, for example, in 1962 at a meeting of the Protestant Workers Association of the CDU/CSU: "Social politics must be responsible for providing the constitution of our society with a formula for integration which neither excludes nor promises to conclusively resolve divergencies, conflicts and differences but which is actively involved in reducing a maximum of friction and demonstrating a realistic basis for mutual interests. If we take this idea seriously, it appears to me that the social market economy boils down to the conviction that it is possible, in the way we have been able to resolve the conflict between social progress and free enterprise in the last decade, to also resolve various other conflicts in different areas of our society. The social market economy is a model for society as a whole."

One cannot deny that the second, the socio-political phase of the social market economy, has not been able to develop anywhere near as much political momentum as the first phase. Indeed, it has never actually become reality because its coordination systems have really been orientated economically and socially too much to be able to provide any room for development in other directions. There is a certain arrogance in regarding an economic model, be it ever so successful, as a general cure-all for the evils of society. In the late seventies there were diverse attempts within the framework of various symposiums and conferences to transpose the social market economy as the ideal model for a new order of international economics for countries in the Third World. (In the meantime, Japan, which no one was paying any special attention to or trying to convert with missionary zeal, was happily going its own way, setting clear national goals.) Those who can remember these events will have to admit that there was an unreasonable attempt at mystification.

Nonetheless, the goals that Müller-Armack set out to specify, seen in a broader context than that of a restricted economic model, remain fundamentally correct. They might even prove to be of political value for us today – not exactly as the "second phase," the crowning superstructure of an otherwise unchanged order, but rather as the first phase and foundation of a new order.

The reconciliation of economy and ecology is not possible in the old industrial and infrastructures. The reconciliation of the welfare state and community solidarity is not possible without a new social policy based on post-industrial social ethics. The reconciliation of technology and the work world is not possible without fundamental revisions in our labor, social, and tax laws and regulations and without changes in our concept of education. And the gradual elimination of conflicts cannot be accomplished in any of these areas without cross connections to other vital complexes and without also considering the respective consequences for the individual.

This is the structural parallelism of technological and social developments which was described above as the third line of development in the transformation of society. Technological and social patterns of organization can – I am expressing this here as a conjecture although it certainly could be verified in historico-cultural terms – be reflected in each other to a certain degree: large-scale technological systems which are characterized to a large extent by a division of labor produce social collectives in which the individual has a relatively low degree of self-determination. Decentralized, network-interrelated technologies find their best social counterpart, on the other hand, in small, flexible, and autonomous groups which allow their members a large share of authentic authority.

Striving for a society of reconciliation can only seriously be undertaken if, aside from observing certain principles in political dealings, one attempts to make adequate social and cultural use of the technological opportunities for development. At the same time it is necessary to constantly keep in mind just how young, in terms of the history of civilization, the conflict between technology and culture is, and to what degree this conflict is tied to a certain type of technology: electromechanics.

"Techne," the Greek word for a manual work of art, stood for many centuries, not for the separation of the outer and the inner world, of man and the environment, but was rather used to denote their symbolic incarnation in masterworks. Without knowing nature, without respecting the unattainable creative power of the Maker, "technological" accomplishments were deemed inconceivable; as late as the late 18th century Schiller's "Glocke" (The Bell) conveys a powerful impression of this unquestionable reliance of technical ability on nature and "religio." The mechanistic concept of nature by Bacon, Descartes and Leibniz, which people today like to regard as the philosophical beginnings of an alienation from nature, were also understood in their day as a triumph in the recognition of the similarity in nature betwen human and divine or natural creative power.

The cultural pessimism of modern times is based mainly on the assumption that nothing will essentially change in the quality of existing technological and industrial structures. There are no other alternatives therefore to the negative aspects of civilization – the overexploitation of nature, alienated work, a non-communicative specialization, run-away consumer buying habits, a drifting apart of the generations – than radical solutions: to deny, to drop out of society, to try and escape into romanticized utopias.

But technology is in the process of a fundamental and qualitative transformation. An inexorable process has been set in motion by which technology, nature's onetime relentless enemy, is becoming one of the environment's most active protectors and even a (biogenetically controlled) natural production factor. Its intellectual, even its aesthetic appeal is continually growing; once the – actual or supposed – antipode of the human intellect, technology is gradually becoming man's "intellectual partner." The more sovereign, creative, and unorthodox man is in playing his role, the more successful this partnership will be. Those who have specialized in simply acquiring factual information will discover their most dangerous rival in the computer while those who have a creative bent and are imaginative will recognize the computer as a most valuable assistant in that it relieves them of routine work. Striving to regain integral interrelationships and an integrative way of looking at things involves, perforce, having a positive attitude towards the co-existence of rationality and irrationality, work and fun, the thirst for scientific knowledge and the desire to express oneself artistically, in technology and culture. A society of reconciliation will be an immanently technological society of culture or it will be less reconciling and, as I am afraid, have less worthwhile cultural elements to show for itself.

One more aspect must be mentioned: Only a society of reconciliation can also be a truly open society. It is by no means a matter of chance that in the Federal Republic of Germany, but also in other Western European countries, a collective preoccupation with oneself was most rampant when the political influence of conflict theories and delimitation strategies was at its peak.

Theoretical approaches of conflict force energies inwards. They direct attention to one's own state of mind and condition. They cultivate narcissism and self-opinionatedness, they underestimate outside forces and neglect the tomorrow. In a society of conflict the tone is fault-finding and lachrymose, the philosophy is scholastic and hairsplitting, the culture is strong in convictions and weak in comportment, the politics are big on words and short on deeds. There is a petty-mindedness and provinciality attached to it, a sense of constant dissatisfaction, a kind of taking offence on principle.

On the basis of its internal make-up, this kind of society cannot be cosmopolitan and therefore does not not pay much attention to what is happening all around it – with the result that the world, after initially shaking its head, likewise starts to ignore it. Those who have the opportunity of frequently discussing matters with nations in the Pacific area or the USA know to what extent strides in critically outdistancing Europe's current standard of culture have already been taken in such countries.

Cosmopolitan attitudes, in all likelihood, though, will be the most important prerequisites for the self-assertion of nations in the upcoming information technology age. Economic success in the long-term is destined for those who are capable of showing initiative in branching out into foreign markets and sympathetically understanding the culture, traditions and mentality of other peoples. There are only a few German products left which are of such high quality that they are without any kind of competition and can be sold, so to speak, without budging from one's office desk.

The question has long since been decided in the scientific field: if you do not publish in English, you go unheard internationally. The standard of achievement in research is generally regarded as the most reliable criteria for determining the future strength of a country in that the period for realizing product investments is now generally very short-lived, so that being ahead of the research game today can be directly projected onto tomorrow's share of the international market. (The fact that this correlation cannot automatically be established for Western Europe because there is not a uniform economic market and frequently a lack of the necessary production capacities for competing on the international market, is another matter; in the eyes of our most important competitors, the USA and Japan, research in any event is given the status of being an "economic barometer.") The internationality of research, particularly in pure and application orientated basic research, makes the lack of cosmopolitan attitudes in one's own country a decisive factor in slowing down development: The best qualified scientists solve the discrepancy by "voting with their feet," by emigrating, in other words. The last decade has witnessed an exodus of German natural scientists to the USA, and we will be suffering from the consequences for a long time to come.

Finally, culturally the presence or absence of cosmopolitan attitudes marks the dividing line between a participating, vital, and a receptive, museum culture. A number of non-European economic leaders and politicians have already assigned Europeans, for the future history of civilization, the "noble" role of being the proper curators of a great cultural past. As a Dutch businessman expressed it at a management forum in Vienna in May of 1985, Japan seems to allocate world roles in the year 2000 as follows: "The United States supplies the food, Japan the industrial products, and Europe is the world's discotheque and culture museum."

Modern cultural impulses, in other words, will no longer be accepted internationally unless they are embedded in scientific, economic, and technological strength. The mastery of modern technology is now seen internationally as an expression of the cultural, and not just the economic vitality, of nations. The reconciliation of technology and culture, overcoming the specific German hereditary burden of "two cultures," is therefore urgently called for to save at least the *one* culture from being condemned to insignificance.

This is why technological and cultural policy belong inseparably together. The one without the other, depending on how they are weighed, would give rise to either technocratic or idyllic narrow-mindedness. Top quality and widespread support are political terms which are just as legitimate in the field of science and technology as they are in the area of art and music. This fact signifies a great opportunity, but also a serious responsibility for German sponsorship measures. It is up to the different states in the Federal Republic of Germany, (that are invested with the constitutional authority, either solely or in auxiliary capacities, to decide policy in matters relating to higher education, the promotion of culture, school education as well as technology and infrastructure) whether the tracks will be laid in the direction of cultural integration or confrontation. Barely a decade has passed since the breakdown of the confrontation model and we are still coping with some of the disastrous aftereffects of the attempt to ennoble certain political science and sociologically orientated branches of culture. This should be warning enough in the future to decisively counteract any attempts at cultural hegemony – whether they originate in research laboratories or social science seminars.

The burden of our recent history and a divided Germany will make it difficult for a long time to come, if not impossible, to express our cultural goals as national challenges. A proclamation like the Japanese government's of national goals to be strived for in the 21st century, "Japanese souls – Japanese technology," is virtually inconceivable for our country. The controversy regarding nationalism that would immediately flare up would end up nipping any kind of future dynamics in the bud. This does not have to mean, though, that we should entirely sacrifice the motivating force of models and projections for stirring up community spirit. Building on a society of reconciliation would be an ideal postulate which could lead to satisfactory results inwards (as a principle in social and cultural dealings between the generations) and outwards (above all in our relations with the German Democratic Republic and other Eastern Block countries). To keep European culture alive with the vitality of European technology, to guarantee Europe's identity of cultural diversity in the next century, this seems to me the most convincing maxim for a new process of integration.

However, in politics, as in everything else, it is easier to form worthwhile resolutions than to carry them out in practice. It is not always the lack of funds which sounds the deathknell for many projects, many ideas end up getting strangled by the red-tape involved in decision-making processes. Approaching election dates leads to putting off and not resolving conflicts; public pressure from the media and interest groups force short-sighted decisions to be taken. Originally very clear-cut systems end up gradually being transformed into casuistic piecemeal work.

To make the idea of reconciliation pragmatic, therefore, it is absolutely necessary to see it not just as an opportunity but also as a duty. It is not an arbitrary, procedural option to be brought into play as one sees fits, a matter of magnanimity or political climate. It is a self-disciplining control instrument, an intellectual guiding principle, an incorruptible authority of control and self-control.

Being serious in striving towards reconciliation as a political principle involves, in the cold light of day, much more discipline than any other form of political action. The consequences are as follows:

- the absolute need for information and communication because the different standpoints have to be clarified and one's own position cannot be considered as absolute from the start;
- the absolute need for cooperation because the consensus reached has to be as broad as possible;
- the absolute need for flexibility because insights based on new information processes and developments have to be taken into account in the course of arriving at a consensus;

- the absolute need for long-term planning because the solutions that are found can only claim to be reconciliating in character if they can prove their durability over a certain period of time;
- the absolute need to reach decisions because conflicts cannot be resolved by eventualities but only by reality.

Since political measures can have very different consequences, these criteria should not be taken in the sense of being a general plan. Day to day business does not require a fundamental codex but alert political operations and tightly organized administrative implementation. There will also always be isolated cases which have to be dealt with promptly, bypassing the intensive decision-making process otherwise called for. Generally one can say, though, that the more community orientated the questions are and the more far-reaching their consequences are in terms of time and content, the more urgent the need there is to subject the matter to this sequence of procedural steps and the examination of its essential content.

One quickly discovers that conventional methods and means are often inadequate for obtaining such results. Particularly the needs for flexibility and long-term planning are often treated in practice as alternatives which necessarily cancel each other out. The result is short-sighted manoeuvring instead of suitable flexibility and that long-term planning tends to be characterized by inflexible systems instead of by true-to-life ranges of variation.

Both politically and administratively, there is still an endless amount of work to be done. The wider the currents of information flow, the more complex the matters become and the quicker politically relevant data and facts change the more danger there is that the priority of politics over administration disappears, and the realities of life and the needs of the future are not adequately taken into account in the political decision-making process. Administrative structures, but also the way in which the different constitutional organs conceive their own and the roles of the others, are responsible for erecting many hurdles. Reforms are imperative, however: Citizens can hardly be expected to be flexible in terms of their work and private life, to be communicative and make provisions for the future while the "good old government" does not change at all.

Changes ought to be made first in the way politics sees itself and then to filter down into the different executive levels. It would be a good idea, for example, to restrict laws or regulations dealing with matters for which long-term predictions cannot be made to certain limited periods of time. It would also be advisable to assign really long-term topics to commissions which are politically acyclical, that is, which continue in existence beyond the end of a legislation period. This is the only means by which such work can be carried over into the next legislation period intact and not be flogged to death in election battles. The public budget should draw a stronger line between investments for the future and current spending. To start with, further net credit lines could be valid when dealing with expenditures for research support while current and consumer orientated spending would have to be financed without going further, or at the most, minimally into debt. The current budget system basically operates along lines that run counter to this type of debt splitting, namely that the net sum of revenues has to be used to cover the net sum of debts. The principle of "non-affectation", however, is already being sacrificed and for far less important causes – for example by the levying of quasi-taxes for a specific purpose. It hardly seems justified, therefore, to stick to this principle when it is a matter of an adequate fiscal differentiation between politically relevant future and current spending.

The necessity of politics to be better informed and to unite the different interests of society in a viable consensus is constantly growing. This urgently calls for new political leadership and communication structures. The ministerial administration must be enabled, with the help of information technology network systems, to comprehend complex facts that overlap different departments and to derive from them "leadership orientated" information. Politics would be well-advised, wherever possible, to discuss these alternatives in direct dialogue with the groups and organizations affected and to make this the test bed of their suitability, and to what extent they represent a consensus. One could raise the objection that this is already being done. The truth of the matter, though, is that as a rule such talks only take place when hearings are laid down by law and these occur almost without exception on the administrative level.

It would be taking things too far to go into further details and regulations. There is a certain justification, however, in maintaining the standpoint that what one is basically dealing with here is simply the question of efficiency and not a political position. On the other hand, there is no denying that strategies for overcoming conflict can only be successful if they are more than just a matter of the mind's eye. They have to be concrete programs with both feet on the ground. This is most likely to be the case when politics knows how to combine a maximum of information and knowledge with an optimum of freedom of choice and decision. Intellectual independence should not be allowed to run up against traditional structural hurdles before it has barely left the starting line.

Reconciliation as the central idea for an informed and integrated freedom – wanted with an eye to the future, supported by policy, and put in concrete form by results – would be a political element of culture that could help considerably in easing the period of transition from the old to the new structures. The result might not be fewer, but perhaps more worthwhile and fruitful fights in day to day politics – which would be achieving a great deal. The experience of rediscovering that a nation and its political representatives could also work together in times of crisis towards orientating themselves to the future, would likewise lay the foundation for the success of the society of the future. Whether it will be called the information society, the society of reconciliation or something entirely different is not important. As far as man is willing and able, it could in any event set up the important political, economic, technological, social and cultural signposts on the road to the next century which the citizens of Europe are now afraid to look down. This alone counts.

Europe: Little Big Continent

Chapter 18

Somehow Too Big and Too Small

What is Europe's Real Position?

If it is true what people say, that persons who have been declared dead have especially long lives, then Europe need not have any fears. Since 1982 at least the media have been more or less lovingly making the funeral arrangements for the Old World.

"The end of our future" was announced more than two years ago by Bruce Nußbaum, the American co-editor of the venerated magazine *Business Week*, by which he meant Europe's, and in particular West Germany's future. The Federal Republic of Germany is like "a car that has gone off the far end of the pier and sort of freezes for a moment in time before plunging into the sea, blind and unconscious of the economic catastrophe which has already overtaken it" in that it has not been able to make the transition from the technology of the 19th to that of the 21st century. In a speech before top managers in 1982 a German ambassador saw Europe sinking into insignificance. A German economic journal even recalled a remark made by William Henry Seward, the American secretary of state at the time of Abraham Lincoln: "The Mediterranean is the sea of the past, the Atlantic the sea of the present, the Pacific the sea of the future."

The president of the World Bank, Alden Winship Clausen, also gave a speech in Tokyo in early 1982 in which he made no bones about "Europe's momentum having come to a standstill" in his opinion. Not long after that the *Frankfurter Allgemeine* quoted the minister president of Malaysia, Mahathir Mohamad, as saying: "For a long time we looked to the West, just as Japan did in the early stages of its development. But the West is no longer a suitable model. It has lost its dynamism. The people there want a comfortable life without facing up to the realities of the international market which they can no longer dominate. If we try to imitate them, we will end up going under with them. The Japanese, on the other hand, still have a stormy development ahead of them. That is why they are a much better model for Malaysia."

It was not long before a term was found for the sickness which was affecting the Old World: Eurosclerosis. What is meant by this was defined by the *Europa-Archiv*, a publication of the German Society for Foreign Policy: "Technologically outclassed, comfortable, decadent, inward-looking, protectionist, conservative."

Despite these dressing-downs one could still cherish the hope for awhile of having something to say, if not technologically, at least in the area of finance. These illusions were shattered again by *Business Week* in 1984: "New York has become the gateway of the international financial system; it has brought the European dominance of the world financial market, which has lasted 300 years, to an end," this Wall Street weekly categorically declared.

The shock for the Europeans following this dramatic swing in public opinion by a world which used to look up with respect to the continent which brought forth Western culture and technological civilization, hit (and is still hitting) home. At first there were only whispered mutterings of disapproval which, however, gradually become more audible. Business associations, particularly those in the field of mechanical engineering and the chemical industry, pointed out that they still had a large share of the world market and growing export rates. The European telecommunications industry could claim, in any event, more than a 26 per cent share in the worldwide production of modern communication technology, in comparison to 39 per cent for the USA and just eleven per cent for Japan. The manufacturers of robots were able to report that there were already more industrial robots in Western Europe than in the United States. They also indicated that Japan's reports about the fantastic number of robots in operation had to be taken with a grain of salt since a lot of what they termed robots would have been classified here simply as numerically controlled machines.

The statisticians compared the share of research expenditures in proportion to the gross domestic product and ascertained that the Federal Republic of Germany with 2.5 per cent (in the meantime it has risen to 2.8 per cent) was on par with the USA and ahead of Japan. Although the December 1983 issue of *Der Spiegel* jolted readers out of their pre-Christmas mood with its title story: "Verschlafen wir die Zukunft?" (Are we Sleeping away the Future?), it also immediately sounded the all-clear: "There is no reason for a feeling of doom."

We are not doomed, but there certainly was and is cause for concern. The European Community countries, almost simultaneously and without exception, began displaying a variety of scientific and technology orientated activities. They were directed, with different emphasis, towards the same identical goals: the promotion of micro-electronics, opto-electronics, laser technology, sensorics, production automation, telecommunications, and biotechnology. Great Britain established a separate ministry for information technology, France launched a five-year-plan for the reindustrialization of the economy, the German federal government passed a "Government Report for Information Technology" which provided for one and a half billion dollars in federal funds for development and research in the new information and communication technologies.

The European Community likewise became active: The research program of the Forecasting and Assessment in Science and Technology (Fast) group was presented in December 1982. A description of joint long-term research and development goals for the European Community was included for the first time. Preparations began at around the same time for the Esprit program which was then launched in February of 1984. The European council also initiated six other research programs in the same year, among them a program for the utilization of modern technology in industry, a research action program in the field of biotechnology as well as a 750 million dollar funded program for nuclear fusion research and a plan for stimulating scientific and technological cooperation in Europe. Furthermore, the European Council at its conference in Dublin in December of 1984 requested the European Commission to report on "measures for reinforcing the technological foundation and the competitiveness of the Community's industry." The Commission has presented its findings in the meantime, along with a comprehensive and thorough report on the perspectives of modern production technology for European industry.

Even a doomsday prophet like Bruce Nußbaum could not help being impressed by so much energy. In a speech held in Frankfurt in October 1984 he conceded, despite some lingering doubts, that Europe at any rate had woken up and was facing the future once again. In his opinion American businessmen and journalists were two years behind the times in the way they perceived Europe. Two years, as one is now aware, is an eternity in the course of development of modern technologies.

More than the Americans and the Asians, though, Europeans themselves seem to be impressed by the spirit of progress which has spread like wildfire from the North Sea harbors to the Mediterranean coast. They look to the success of Ariane and Airbus, European scientists on board the Spacelab research laboratory jammed packed with European technology, and the exorbitant rate of increase in exports, especially to the American market (estimates are that exports to the United States easily accounted for half of Western Europe's growth in 1984). At the same time our Atlantic partner had a record trade deficit of 123 billion dollars in 1984, and there were alarming reports about serious drops in sales and about bankruptcies in the American electronic industry. Some politicians and managers interpreted this development as a sign that everything was all right again with the Old

World. Anyone who closely followed the fierce debate concerning the SDI research program could even discover that some people were not entirely free of a certain European arrogance, according to the motto: Without our know-how the USA will not be able to make it on their own anyway.

Caution is urgently recommended here. In contrast to the Latin saying, it is not enough to simply want to achieve great works, one also has to put them into practice in today's international competition for positions of economic leadership. At the eleventh hour – and this can be regarded as a historical achievement – Europeans have realized that they were dangerously on the edge of a precipice. They are no longer marching blindly in the wrong direction. We are not out of the woods, by any means, though; to assume this would be just as presumptuous and short-sighted as thinking that overcapacities in the micro-computer branch indicate a structural crisis of the large competitor nations. New markets are always dynamic markets: The fantastic sales and profits achieved by the pioneers of the industry act as a magnet for other new businesses which want to have a share in the action. This is especially true in the United States where getting ahead ranks with baseball as one of the nations's favorite sports, winners are admired and popular, and venture capital can be had over the counter. This leads to hard-fought price wars with some businesses falling by the wayside in the process. The net result, though, is a solid new industry infrastructure which is equally productive and competitive and which ends up producing a lot of young complementary businesses and new jobs.

What is Europe's real position today? This question cannot be answered either in simple or straight-forward terms. Europe's situation is complicated because Europe's legal and actual status is complicated. There is the unbalanced, the double-faced, and the paradoxical Europe; a homogeneous, forward-looking Europe acting decisively and rationally is still in the making.

Our continent is neither fish, flesh, nor fowl in cases where it acts neither consistently as a unit, Europe, nor consistently as separate nations, as non-Europe. European satellite technology, for example, can bear comparison with international standards. Things have worked out very smoothly so far in the partially bilateral (TV-SAT) and partially European (ESC) collaboration. In Ariane meantime we have a booster which has become a real rival for the United States. On the other hand, the postal authorities of the different nations were not able to come to an understanding concerning television standards for some time. A regular post-industrial war raged for years between PAL (The Federal Republic of Germany), SECAM (France) and D2-MAC (Great Britain). It was not until the summer of 1985 that agreement was reached on the British color television standard.

Altogether eleven different digital broadcasting technology systems, costing more than seven billion dollars, were developed by nations in the European Community; for the much larger market of the United States and Canada there were only four. Provision for interfacing the systems was not achieved until most of the development work had been brought to a conclusion in the countries involved, likewise in mid-1985.

The specialists are fully in agreement on the importance of standardized modules for the manufacture and use of automated production systems. In the United States hundreds of American companies, among them General Motors, Ford, Chrysler, IBM, and General Electric have reached agreement on a standardized interfacing system for the production facilities. In Japan twenty large corporations are working in the research city of Tsukuba on a standardization of modules for production facilities working with laser technology. Excerpts from the report of the EC Commission for Modern Production Technology in Europe tell a different story: "Although Europe has considerable know-how at its disposal, the Europeans have fallen behind the enormous efforts of the United States and Japan. The commission was not able to achieve a standardization of interfacing systems in its talks with the relevant organizations of the machine tool industry. This was due above all to the fact that a large number of those concerned, who are responsible for very different businesses with different technological standards in the conventional sector of the mechanical engineering industry, have not yet recognized the importance of this problem for the most part."

Given this situation improvements can most likely be expected by direct cooperation between countries and companies. In order not to entirely miss the boat when megabit chips swamp the international market in a year or two, a German and a Dutch industrial giant have been doing joint research in this field since 1984. The estimated development costs of two billion dollars, of which the German Federal Department of Research and Technology and the Dutch Department of Economy are contributing 300 million dollars, drastically demonstrate the dimensions and risks involved in the area of micro-electronic mass production. A different joint project between German and French computer manufacturers which was under proposal fell through. In the meantime, however, a German-French Strategic Forum for Micro-electronics and Computer Science is endeavouring to coordinate industrial research and development capacities across the borders. The universities of Karlsruhe and Straßburg recently worked out an agreement for a joint research program dealing with automation and robotry in which a fibre glass line will provide for the exchange of data between the two universities.

In most areas, however, that have not been directly affected by the pressure of competition from high technology like micro-electronics there is a strong tendency to work in "splendid isolation." This attitude is encouraged by public acquisition markets which are still dominated by national interests and the number of legal obstacles put in the way of joint efforts between European businesses. The more progressive the technologies, the more protectionist the government's policy is towards inviting tenders and allocating projects. A notable exception to this rule, which is brought up all the time just because it is so exceptionable, is the agreement between the German and French governments for a joint invitation of tender for the development of a mobile digital wireless telephone; it looks as if a reasonably priced car phone system for Europe will thus be realized within the decade.

Joint efforts between companies run up against various legal problems which are the more serious the smaller the economic partners are. A legal framework for the "European Company," which has been on the Community's agenda for more than twenty years, still has not been established. The consequence is that if a German and a French company want to work together on a joint project they have to found a new company which falls within the jurisdiction of either German or French laws. The different national laws concerning corporation taxation are no more in line with each other than standards and technological regulations. The European Community's regulations governing competition also subject research and development agreements between companies of different nationalities to the very stringent restrictions of the EEC's anti-trust laws. After more than ten years of negotiations so-called group exemptions concerning patent license agreements at least have been made possible. On the other hand, in the area of development or even sales the matter "cannot yet be considered ready for passage," as the German federal government announced in their report dealing with integration in the European Community in late 1984.

The double-faced Europe is closely related to the unbalanced Europe: Since it often indecisively vacillates between old and new structures, it looks forward and backward at the same time, wanting to open up to the (information technology) future without making a clean break with the (traditional-industrial) past. Many examples for this kind of inconsistency can be found in the work of the EC committees. Trying to please all twelve partners all the time ends up producing a lot of this-as-well-as-that instead of clear either-or decisions. The same committee within the short period of a few months can speak out for doubling the research expenditures of the Community and for extending the ship building subsidies, thus voting for and against making structural adaptations.

In its work program for 1985 the EC commission promises "an environmental policy with European dimensions." With a certain pedanticism it then goes on to add the following: "The necessity of an environmental policy is nowadays indisputable. Nonetheless, its importance is often underestimated in comparison to other dimensions of economic and social policy. This is probably due to the fact that environmental protection can result directly in inconveniences and costs while its positive consequences are only felt in the medium or long-range. The commission for its part wants to award environmental policy a strategic dimension within its economic and social policy of orientation. It is not just a matter of making up for the damages which were done in a phase of industrial development that was without any consideration for ecological interests. It is also a matter of anticipating and speeding up an industrial change in which environmentally relevant factors are gradually included."

When it came to setting a deadline for the manufacturing of less polluting cars on a European wide basis and establishing standardized exhaust fume limits, however, the commission was more impressed by the massive resistence of some member countries than by the necessity of the changes. This aspect of environmental policy can only be considered "strategic in dimension" if one equates "strategic" with "long-range in effect." It will be interesting to see just how long it will take the EC to carry out its plan for the elimination of steel subsidies, originally set for the end of 1985 by the Steel Subsidy Code. Whether steps toward the "major goal," described in the commission's work program as "the creation of infrastructures for the post-industrial society," will be taken or whether, and several factors point in this direction, steel subsidies will be extended or even increased is a matter for speculation.

Without a doubt, though, EC-Europe is handicapped more than anything else by what one could call its structural paradoxes: The highest degree of reglementation and of levelling out prevails where the structures are most differentiated – in the agricultural sector. Conversely, there is the greatest lack of cooperation in dealing with matters that can afford it the least – in the area of joint basic research, common energy supplies, and standardized industrial norms. Europe is somehow too small and too big: too small to cultivate the illusion of national self-sufficiency regarding technology and energy policy; too big to accommodate the competitive fishing, wine-growing, dairy, grain production and tomato growing interests between Denmark and Sicily, Great Britain and Greece, the Federal Republic of Germany and Portugal for ever.

Although agriculture, more than any other branch of production, is characterized by natural differences in region, climate and ecological make-up, the matter is centrally controlled and authoritatively administered in Europe. As if we were still struggling with the problem of feeding our population, as in the sixties, every ton of grain and every quart of milk, whether from a small farm in the Alps or a modern "farm factory," is subsidized with the same supporting measures to start off with.

The Community now harvests more wheat than the USA, and in the case of most agricultural foodstuffs produces far more than it needs (50 per cent more wheat, 48 per cent more butter, 25 per cent more milk, 38 per cent more sugar, 32 per cent more wine, respective averages of recent years). Nonetheless, Brussel's agricultural policy is solely production orientated and directed towards the expansion and intensification of agricultural products. Basically things function according to the simple formula, if the farmer wants to earn more he has to produce more.

Since everyone always likes making a little more money, the farm businesses of Europe are being expanded, rationalized, bought, leased, over-fertilized and overused. Everyone is behaving in an economically rational manner in accordance with the pace set by Brussels. Many farmers are up to their necks in debts; and because there are no market mechanisms but simply guaranteed prices the early warning signal for surplus production – falling prices – does not function. By the time farmers become aware of the danger, in the form of quantity limits (contingents) and quota regulations, it is of course too late. Most long-range investments have already been made and the farmers end up feeling that their political trust has been betrayed and that they have been cheated of the fruits of their business enterprise.

Nothing has had a more lasting negative influence on Europe's image, both externally and internally, in recent years than the permanent quarrels within the European Community concerning agricultural and financial policy. The European spirit has been torn in half by table wine and fishing quotas, the summit meetings of European heads of states turned out time and again to be torturing ostracisms of European politics turned upside down. Whenever the representatives of 270 million people have met in recent years, not perspectives but prices, not progress but finance were the central themes of their discussions.

Agricultural policy developed into a measure for Europe's state of condition although this was all out of proportion to its numerical importance and its share in contributing to the total European production and job capacities. It was even more incongruous in view of the urgent structural problems Europe was actually facing. Europe's international competitors could not have found a better advertising agent than the Community's estranged agricultural and administrative policy. An accident of history, compromising on the lowest common denominator of an integrated agricultural market after other more far-reaching plans had fallen through (European defence community, European federal states), is threatening to become a historical inevitability of a dissipating and disintegrating continent. The paradoxical Europe is the Europe without a technological community and without a common domestic market, the Europe of ecological and regional neglect, the Europe in which inequalities are treated in an identical manner and common interests are artificially segregated.

Europe's real position is, therefore, more a question of attitudes and institutions and only in the second instance a question of economy and technology. The Community will soon have to decide whether it is going to give precedence to an improved convergence between research, technology, economic, and monetary policy of its member countries or whether it wants to leave things the way they are now with the predominance of agriculture. At the present time it still thinks it will be able to combine the two. This is laboring under a misapprehension, however.

As long as the common agricultural policy gobbles up more than two thirds of the total budget, it will continue to dominate EC policy themes while other laudable future orientated plans can make do with gathering up the crumbs. The research programs established by the European Commission in 1984 typically enough were only resolved on for a period of two years to start with. After the exorbitant agricultural bills have been paid off, there is virtually just a pittance left over for the financial research support of other projects. In the 1984 Integration Report of the federal government of Germany previously referred to one can read: "The Council took up the question of the priorities of joint research at their meeting on June 29th, 1984 ... The German and the British delegation emphasized that, in view of the well-known financial constraints, it was questionable whether all the proposed projects of the Commission could be financed. The budget situation demands, aside from a time schedule for the implementation of the proposed projects favored by the other member nations and the Commission, making selections among the projects themselves. The Council was not able to set concrete priorities for such a selection."

Financial support of research perspectives in the USA and Japan is provided for a period of ten or even twenty years – the Japanese program for the development of the fifth generation of computers alone extends over a period of more than ten years. In comparison, the European way of going about things is like trying to make progress in the area of nuclear fission using Otto Hahn's laboratory techniques. Research policy which is not designed with the strategic dimensions of time and goals in mind, which are the determining factors of its financial framework, is not competitive by international standards.

As long as the European Economic Community is in fact nothing more than an European agricultural common market, it will scarcely be in a position to establish a coherent and dynamic foreign trade policy. Its political force is confined for the most part to permanently trying to reconcile the divergent agricultural interests between its members. The sources of friction are numerous, the disagreements manifold, the danger of inward-looking is obvious. Outwardly, there is often just enough energy left to erect protectionist barricades; for a concentrated and concerted advance into the dynamically growing non-European markets, in contrast, the force is missing.

More than half of the EC exports are to other countries in the Community. If one considers Western Europe as a common domestic market, which corresponds after all to Europe's own self-declared image, one is forced to considerably qualify the export success of the continental industrial nations. This is even truer if one keeps in mind that the current favorable rate of exchange is for most part responsible for the expansion of exports to the USA giving these exports a highly temporary nature. Our own imports from threshold countries, with which European businesses enlarge their range of products, have also increased considerably so that a lot of what shows up in the statistics as exports has not been produced by employees in our part of the world.

It was not until 1983 that European suppliers were able to record sale increases in East and Southeast Asia, currently the toughest markets for price-performance ratios. In fact, they had considerable losses in some areas. Exports to the Pacific region have consolidated in the meantime, but we have not been able to profit much from the rapid growth in this region till now. This is due to the fact that our share in high technology export goods, which the Southeast Asian nations are particularly interested in, is still quite small.

Even the Federal Republic of Germany, which has been able to maintain a comparatively good position, cannot rest on its laurels. Exports to the ASEAN countries make up just seven per cent of its total volume of foreign trade, three per cent less than exports to the Netherlands. The eight Asian threshold countries get half their industrial imports from Japan, about a quarter from the United States and just barely twenty per cent from the European Community. It is also a well-known fact that since the beginning of the decade US trade with its Pacific neighbours is larger than its volume of trade with Western Europe. It is one of the ironies of world history that the former developing nations of Asia are now among the most consistent crusaders for free world trade while Europe, which used to support this position with spirit against the demands of the Third World for a new order of world economy, is now considered a specialist in cooking up inofficial trade obstacles.

The German Industry and Trade Association, which conducted a survey in 1981 of around 10,000 businesses concerning their experiences trading with other EC countries, reached the following sobering conclusion: "The number of trade obstacles and their unpredictable consequences for the individual have increased the wish to return to the system of protective duties which is easily comprehensible, not as difficult to deal with, and whose effects are predictable. Thus the credibility and trust in the EC are at stake whose main goal is the realization of the free exchange of goods."

Agricultural Europe is a Europe of specialists and bureaucrats. In 1981 alone more than 1,500 rulings concerning regulations for the grain and rice market were produced, as one can read in a report by the "Kronberger Circle" (Klaus Peter Krause: "Für eine neue Agrarordnung" (For a New Agricultural Order)). This certainly keeps the administration busy, but it does not provide impetus for developing the European integration process. There are basically two fundamental and by no means congruent interest groups within the EC: One group is interested in improving the economic and social conditions for their rural populations with the help of the common agricultural market; the development of a high technology common market, as long as a higher standard of living has not been achieved, is of secondary importance. The other group perceives the concentration of technological forces as the most important and decisive factor in determining future progress in Europe; the agricultural market in their eyes is a hindrance rather than a help in achieving their goals.

The actual situation of the Community speaks for the "agricultural European:" the big chunk – agriculture – takes the lead in the budget of the Common Market's social expenditures (1984: around 20 billion dollars) over all the other areas in terms of integration; its established and extended administrations; the by now rudimentariness of existing national leglislation and executive authorities; and more recently the boost for the agricultural faction with Spain and Portugal joining up. The integrated Mediterranean program, which was established in 1984 to provide support for Greece, large regions of Italy and the southern Departéments of France places an undue financial strain on the Community's budget and is manifest proof of the political power of the Southern European agricultural lobby. Spanish and Portuguese demands to establish similar aid programs will not be long in coming.

On the other hand, a number of factors speak in favor, of the "technological European:" the realities of world trade, the close interrelationship of industrial markets, the interdependency between science and technology and the international challenge in competition which is particularly directed at Europe as a sales market. Outdated materialisticly orientated and institutionalized structures, on the one hand, and the demands of the future for adaptation, effort, and even sacrifice, on the other, are the same essential elements of the conflict which we are facing not only as a nation but also as member of the European Community.

Synthesizing these contradictory positions into a whole is extremely difficult, as the political practice has shown. It certainly does not seem to be a lost cause either – provided that political reason can win over bureaucratic ambition, and historical perceptiveness can overcome administrative inertia. Europe's response to the impetuous challenge of younger, success hungry nations can be neither centralistic uniformization nor strength sapping fragmentation. A third, particularly European way, in fact, must be found which is capable of a flexible and pragmatic coordination of developments that the individual nations cannot cope with entirely on their own, as things stand. As long as the gap between economic and social structures in Western Europe is so broad that one can speak without exaggeration about olive versus technological regions, as long as the pressure to change being put on Europe from the outside is so intense that it is a question of the survival of the continent, as long as this holds true, it is infinitely better to work together to develop functional and project orientated criterias of efficiency (whether bilateral or multilateral) rather than to strive for superficial institutional unity.

To acknowledge this is by no means a contradiction of the European concept of unification which is an absolute historical imperative against the background of the worldwide information and communications networks, and in view of the secular shifts in balance from military to power blocks orientated towards the development of economy and technology. But a Europe conceived according to the convoy principle (whereby the slowest ship sets the pace for the whole convoy) cannot be our goal either because that would be highly dangerous. In the end the world of the old continent will be measured by the force of its revitilization and not by its talent for reglementation.

Chapter 19

Focusing Our Forces on the Facts

The Many Faces of Europe

A number of terms have cropped up recently which reflect the obvious divergency between national economic and institutional structures and those of the Community. They basically all boil down to the one and the same thing: What we need, they say, is: "a two-speed Europe," "a Europe with variable geometry," or the return to the "heart of Europe."

The underlying concept is certainly correct. The alternative of either placing the partners all under obligation to act jointly, if possible on the basis of unanimous decisions, or refraining entirely from political agreements for lack of being able to reach a consensus, is not viable in the long-run. With a disproportionately high amount of friction, only minimum results are often achieved with which hardly anyone is really satisfied. Furthermore, "paymaster" myths, solo efforts, open or concealed violations, blockade attempts and acts of defiance tend to be provoked. Marching in step or standstill: both are out of character for Europe – the one because it ignores Europe's rich diversity with its deep historical and cultural roots, and the other because it does not do justice to the still vital dynamics of the European peoples.

If one accepts these ideas, however, there cannot be simply just two speeds (by which is meant no doubt, fast or slow) for achieving European development. Every subject complex basically has its own logic of integration which is determined by what has already been achieved, by reacting to outside pressure, by the respective status of the different nations, and many other factors. The success of the European unification policy depends to a large extent on whether the intrinsic differences in cooperation speed are accurately assessed so that the potential of the partner nations is neither underestimated nor overestimated.

The European Monetary System (EMS) took this into consideration and has had good experiences with it. Although the English pound is still not part of the intervention system, France, Italy, Ireland and Belgium still maintain considerable control over capital flow, Italy claims an overproportionate range of fluctuation for the lira, and the Federal Republic of Germany does not as yet permit private trade with the new monetary unit, although in other words there is a very wide range in applications by the different nations, the EMS has passed its test with flying colors. The exchange rates within the Community have become considerably stabilized in comparison to the old "currency snake," demand for the common monetary unit ECU is steadily growing, and progress towards a complete monetary union and its opening up to other countries that do not belong to the EEC (Switzerland and Austria in particular) are conceivable. Flexibility and efficiency do make a good team.

Taking the EMS as a gauge, European forces of integration are currently being underrated in the area of research and development and overtaxed by the common agricultural order. Since there is a close political, financial, and institutional correlation between the two, we should develop coordinated measures for quickly remedying the discrepancies. On the one hand, they must give the participating governments a larger scope of action, and on the other hand, provide for more dynamism in joint development efforts for the future.

Integration should not be confused with trying to bring everyone into line. A reorientation must occur in two central areas of European politics and according to two opposing patterns of order – decentralization and extensiveness on the one, and concentration and intensification on the other side. Europe is best off politically, and one should recall this from time to time, when it manages to achieve a balance between the different national forces, and in the greatest danger when hegemonic tendencies and attempts at arbitrary uniformity get the upperhand. One can hardly suppose that things will be much different in the future.

A reform of the common agricultural market requires facing up to the fact that our situation has changed radically. From not being able to produce enough food to feed our populations, we are now swimming in a surplus. There is an acute danger, as more Southern European countries with their large production capacities join the Common Market, of producing even more surpluses which cannot be absorbed by the consumer market. EC stockpiling eats up around a half a billion dollars in annual interest costs as it is. The 1986 draft budget by the European Commission provides for a 16.5 per cent increase in expenditures, of which 4.5 per cent are earmarked for additional costs in regulating the market for the Iberian Peninsula.

Under these circumstances, further support for investments in the expansion of agricultural intensive production cannot be economically and financially justified. Agricultural policy measures should no longer provide incentives for increasing yield in the future. This is the only way to straighten out the financing system which is a constant source of friction and permanently on the verge of collapse. As long as, as is currently the case, repayments from the EC treasury are more or less tied to agricultural surpluses, the member nations behave "no differently from many citizens who try to get the most out of their health insurance because they want to at least get their money's worth back on their contributions and because, from the standpoint of the individual, this seems to be worthwhile" (so Klaus Peter Krause in the previously cited paper "Für eine neue Agrarordnung").

Accordingly, yield and performance in agriculture should not be measured only in terms of production volume. Product subsidies, which make up the largest share of the support costs, should gradually be eliminated for all foodstuffs. Production orientated public support should be replaced by financial aid, which – in contrast to expenditures for regulating the market – would directly benefit agriculture and could take regional needs better into account.

The work done by individual farmers in landscape and nature conservation is of public benefit and not production orientated. This work, especially if it means sacrificing increases in yield, should be awarded accordingly. Establishing and maintaining balanced areas of land in their natural state, limiting cultivation, and cutting back on extensive production methods, particularly the use of mineral nitrogen, are examples for this type of work. The EC should no longer pursue an agricultural structure-policy which places a disproportionate burden on regional and ecological needs, as has been the case so far. Full-time farmers who cultivate on an intensive basis should not be given priority over part-time farmers who usually work with extensive methods. Brussels needs a less regulated agricultural policy which is limited to providing guidelines and which allows its member nations more freedom in providing direct structural aid and compensations.

A great deal of political firmness and some concessions, financial ones as well, are called for in gradually realizing this concept. Great Britain and the Federal Republic of Germany will be more likely to see in each other their natural allies here than the Southern European countries. The fact, though, that both France and Italy have clearly indicated that they are very interested in establishing a European research and technology community should provide a favorable basis for negotiations if it is duly made clear that a cost-intensive agricultural and industrial policy cannot both be realized together. In any event, the main interest of the Federal Republic of Germany in the near future must be to achieve more political scope of action for maintaining a comprehensive, ecologically tenable and less intensive exploitation of the landscape.

The integration logic in the area of research and technology, as opposed to the agricultural market, urgently calls for amalgamating rapidly and for increasing productivity, for the reason that Western Europe, and in more than just one way, is having the screws put on it.

The most important impetus for the technology push in the USA is now coming from the defense department which has doubled its research budget within a period of just five years; with a 1986 research budget of around 40 billion dollars the Pentagon is financing more than 70 per cent of public and almost 40 per cent of the total research expenditures in North America. Whether this is, as some critics maintain, basically just an expensive detour for achieving civilian technological and economic efficiency, is open to question. In any event, most of the scientific and technological forces are in the hands of the government. With a sense of national vision, super ambitious development goals (SDI requires technologies which are ten to a hundred times superior to the ones we now have) and gilded harnesses, they are off and running to secure the laurel as the best and the brightest nation in the world in the the 21 st century.

It is to be expected that as the money pours in from the Pentagon, the flow of security sensitive information towards Eurpoe will tend to slow down to a trickle. "A directive of the (American) defense department, which was put into effect in November of 1984, calls for non-secret information and know-how also being subject to thorough screening if their research has been financed by the defense department. Oral or visual interaction in connection with scientific interchange and international conferences are expressively put on the same level as establishing dummy companies and other illegal activities" the head of the Federal Institute for the Study of Eastern and International Science, Heinrich Vogel, reports in an essay in the February 1985 issue of the magazine *Außenpolitk*. Vogel points out (which corresponds to experiences of European scientists and a study by the Federal Research Department of Germany) that "to avoid trouble with the main financial backer, the Pentagon" foreigners, to be on the safe side, just are not invited to many American scientific congresses. The US Academy for Science also quite rightly recognizes this as being detrimental for international as well as for domestic research.

Japan's avowed aim is also to achieve an undisputed and unapproachable supremacy in the world, not only in the production of component parts and computers but also in all the other important areas of high technology. Its concept is a civil technological "contrast program" to that of the USA, flanked by a clever licensing and foreign trade policy. Finally, the new Southeast Asian industrial nations, in particular South Korea, Taiwan, Singapore and Hongkong, have specialized in being two steps ahead of their competitors in marketing the rapidly changing innovation cycles without spending much on research themselves and playing the card of their wage cost advantages for all they can get.

Our continent is being confronted by a research and industrial policy challenge of global dimension. Europe plays a star role in the script plans of the Japanese and American micro-electronic and computer electronic companies, an indispensable role as an important market for absorbing the gigantic production capacities. Europe's world market share in the production of semiconductors has sunk to 8.5 per cent; in the world production of micro-chips in 1984 it did not even represent more than 6.7 per cent. The Federal Republic of Germany in particular imports more computer products than anyone else; the German computer branch has only about a one per cent share in the market outside the country, of which easily half is composed for the most part of imported technological components.

After seeing what an easy time non-European chip and computer giants had in successfully taking over the market, it is no wonder other high technology manufacturers are eager to try their luck in Europe where customers have money to spend. According to a survey by the *Financial Times*, the industry in EC countries will spend more than 50 billion dollars for automating production by 1990, above all in the form of numerically controlled machine tools. The telecommunications market, with its need for fibre glass and digitalization, will be on an even much larger order of magnitude. For biotechnological products, whose world market volume now amounts to about 20 billion dollars, annual growth rates of eight per cent are considered likely. The scientific journal *Biotechnology News* estimates the market value for products manufactured with gene technology at 145 billion dollars by the year 2000; other predictions are much higher. Whether it will be more or less, whether progress will be made somewhat quicker or slower, is relatively unimportant in all these cases. The fact alone is crucial that Western Europe – the European Community and Efta taken together –, with their world trade volume of over 30 per cent, has a market position which acts as an incentive for its international competitors to exert themselves to the utmost.

Europeans have not only recognized the danger, they are also attempting to react to it. One is certainly not amiss in getting the impression, however, that there is still – this was the case in any event till mid-1985 – a great deal of uncertainty concerning priorities, programs and projects and, above all, suitable levels of cooperation. A whole series of actions are in progress parallel to one another: national programs for promoting research and development in the key areas of technology, multilateral Community projects (particularly in aeronautics, space, and civil nuclear research), technology initiatives by the European Commission, and variable, large-scale projects which are open to any interested countries, such as the ambitious but also very vague Eureka Program.

In addition, there is a growing number of business cooperations and industry interconnections which go beyond the borders of Europe. A selection from among the many examples: American Telephone&Telegraph (AT&T), the American telecommunications giant, formed the AT&T-Philips Telecommunications with Philips (The Netherlands) and acquired interest in Olivetti (Italy); Olivetti in turn has signed marketing and production contracts with Xerox and Toshiba (Japan). Siemens (The Federal Republic of Germany), ICL (Great Britain) and Bull (France) maintain joint research facilities in the area of "artificial intelligence." The four largest telecommunications companies of France, England, Italy and the Federal Republic of Germany are working together on the development of their own new telecommunications components. Nixdorf, with plants in Paderborn, Ireland and Singapore, is negotiating with RCA and LSI Logic (USA) about possible forms of participation in the company's development of semiconductors. These are proofs that the communication and computer know-how of the New and Old Worlds, along with European-wide marketing networks, are joining forces.

Even sales giants have no other choice but to cooperate on an intercontinental basis and to augment their own strengths by making arrangements with their foreign competitors. The technological elements of the information and communication technology, on the one hand, are too diverse to all be able to be produced, and above all developed, by one company alone. On the other hand, the central markets – USA, Japan, and Europe – are increasingly overlapping. They constitute more and more a standardized market potential for high technology business strategies, as the Japanese management consultant, Kenichi Ohmae, recently expounded with great insight in his book "The Power of the Triad." Eighty-five per cent of the worldwide computer production goes to just five countries: America, Japan, the Federal Republic of Germany, Great Britain and France. In the case of numerically controlled machine tools there are just three nations – Japan, the USA, and the Federal Republic of Germany – which account for 70 per cent of worldwide sales. Businesses try their best to be as visible and integrated as their competitors in these enormous "demand pools" with licensing deals, development, production and marketing agreements.

The establishment of a European research and technology community has been a standing item on the agenda of the European Council's meetings since the EC summit meeting in Stuttgart in June of 1983 and a favorite perspective topic of the Commission. In view of developments, its realization for Europe is not only a matter of technology policy, but also a socio-political responsibility. In March of 1985 Brussels published no fewer than three strategy papers dealing with this topic: the Commission's report concerning boosting the technological foundations and competition of the Common Market's industry, the report of the ad hoc committee for institutional questions (the so-called "Dooge Report") and the work program of the European Commission for 1985 which also called for "a strategy for research and innovation". The European Council's unanimous agreement on the Eureka Project at the summit meeting in Milan in June of 1985 was also formed with the intention of "organizing the development and research activities of interested countries," as the French secretary of state, Dumas, said in a letter to his German colleague Genscher.

Is there a danger that the many different strategies will end up cancelling each other out? The idea is certainly not entirely far-fetched. A *political* division of labor in any event seems just as urgent as a scientific and technological one. It should be pursued in strict accordance with criteria of competence and efficiency and make the best use of resources which are already available.

In this scenario the European Community should act as an authority in providing guidelines and general coordination rather than as an organ for dealing with concrete research or technology policy. It cannot and does not have to compete financially on the amount of money that most European industrial nations are now spending on important research purposes. Its administration, which is costly, centralized, trained for regulating details, and subject to various voting requirements, is more of a hindrance than a help for the flexible, application orientated research sector; more than a handful of German middle-sized businesses have thrown in the towel in just trying to fill in the Brussels application forms for the Esprit Program.

The European Commission is indispensable, on the other hand, when it comes to establishing common industry standards and a uniform customs procedure, bringing the different national corporation laws into line with each other, and creating the prerequisites for European-wide transportation and communication infrastructures. Furthermore, it can act in important preparatory and coordinating capacties in opening up and harmonizing market divisions. Things, for example, have already gotten off to a promising start in the area of telecommunications: The European postal administrations, which are organized in the CEPT, reached agreement in early 1984 to open up the formerly isolated telecommunications markets on a European-wide basis. The Community's Race Program (for research and development in the area of modern communication technologies in Europe), in a logical and consistent manner, continued this development in its efforts to establish a standardized European wide-band communications network. After 1995, according to the Council's plans, there should be a standardized, digitalized and service integrating fibre glass network in the Community, implying that the market for many of the auxiliary devices would also be standardized for the most part.

Initiatives and concrete schedules for creating the normative prerequisites for an integrated domestic market, harmonizing the technological infrastructures, coordinating scientific and industrial cooperation between countries as the cases may require, gradually deregulating and regionalizing the agricultural market – the scope of activities and responsibility is sufficient to test the indispensability of functioning EC institutions in the near future. The realization of the goal to actually create a standardized EC Market by 1992, which the Commission has envisaged in its most recent white paper and which was confirmed at the Milan summit meeting of the European Council, alone requires enormous effort. Among other things, indirect taxes have to be brought into line with each other in order to be able to do without customs controls.

One has to warn, however, against more far-reaching eurocentralist ambitions of suppression and substitution; they could end up causing more harm than doing good. Behind the national research programs, which Brussels is increasingly making no bones about seeing as annoying rivals, there is rarely ever a kind of megalomania on the part of any single European nation wanting to take on the world all by itself. These measures rather, in being tailored to different economic and research structures, are indispensable for regional structural adaptations which cannot be controlled on a centralized basis.

In all our efforts to expand the dimensions of the inner-European market, we should not forget, that the standardized economic area we are striving for has to provide room for a great many. Small and medium-sized businesses are the backbone of the national economy in almost all the partner nations, both in terms of economic and employment policy, as well as in terms of technological innovations. Trying to stretch and pull our future plans to fill the Texas-sized shoes of a few non-European information technology giants by concentrating on billion dollar sales and mass production is the surest way of breaking Europe's supple backbone. The development of multi-megachips and super computers can be made an international and consortial research goal, but without being too petty as far as either financing or anti-trust laws go. The overall structure of the European technology community should not be tailor-made though to fit these "mega-dimensions" because Europe's greatest strength lies in its decentralized, creative, and spontaneous diversity. The insects have survived the dinosaurs – one should also think about this from time to time before being so eager to blindly follow in the footsteps of the others.

Focusing forces, yes – but in keeping with practical and then institutional criteria: This seems to me the only promising road towards a European technology future. Logically enough, in designing multilateral research and development projects it is only of minor importance whether the nations which are interested and willing to commit themselves belong to the European Community or not. Eureka is open to any Western European country, and this fact might very well prove to be its best chance for success.

This means of course reversely: the belief that a European technology community can be used as a vehicle for achieving a European union, is also wrong. There is no logic, functioning according to historical laws of its own, either in terms of an agricultural or a research policy, which demands the creation of large-scale political and cultural unities. Wanting to combine both by package dealing will not advance political Europe, but will only be a tremendous handicap for its economic and political renewal.

The worldwide dynamism ensuing from the rapid advances being made in every branch of high technology is putting political decision-making under pressure. Taking pragmatic and rapid action while exploiting existing structures that have proved the test of time (among other things, the joint ventures by European companies and universities as well as existing joint research institutes) is an imperative. Provisional and reversible, from project to project variable forms of cooperation could be quite useful, at least in the initial stages of the technology community. Proven success in limited areas, international program agreements and multilateral research consortia with clearly defined goals are at present politically and psychologically more important than the perfect idea of success which might be a long time in coming. This is even more true in view of the fact that the preliminary project search phase in the SDI plan is much further advanced than that of Eureka, for example. Despite the fact that the one "technology package" is tied up with military ribbons and the other with civilian garters, their research and development goals overlap each other and are even partially identical. Most of the potential European companies are already in contact with the SDI planning group and taking part in concrete talks while they still do not quite know what to think about Eureka. If the structural, organizational and financial preliminary stages of the Eureka project get dragged out too long, there is a danger of the capacities of the best European businesses in the area of research and development being tied up in SDI contracts. The personal and financial resources of Europe's businesses are not up to dealing with two "jobs of the century" at the same time. Eureka has to make haste if it wants to have a chance of succeeding at all. Not structural models, but concrete descriptions of the available contracting jobs are needed so that European companies can put in their bids.

From the standpoint of productivity, the common agricultural market is indisputably an enormous success. Its positive influence on European politics is ambivalent, however, and in terms of social policy it clearly deserves critical assessment. The experiment has shown that partial integration does not necessarily always incorporate an intrinsic logic of continuation in the direction of total integration by any means. On the contrary, they come dangerously close to being "organized absurdities" if one tries to overload them with too much irrelevant model designing. The common domestic market, the research and technology communitiy, and above all the important goal of achieving a unification of Europe which is supported by its peoples and carried out by its governments and parliaments, must be spared a similar fate.

The political Europe cannot be laid down from above or made up of parts; it grows from below, as a state, as a conviction, yes as a feeling. It grows out of the self-evident truth of being, and what politics above all can do and should strive for, is to gradually give this self-evident truth more certainty and steadfastness – by comprehensible deeds and identifiable successes. The Community has already accomplished a great deal on this road; it has also let itself be led again and again into making the mistake of rhetorically overshooting its mark and announcing deadlines which it could not keep. Free peoples do not let themselves be influenced by deadlines and organizations in the process of shaping their political ideas. They have to want change, when and because the conditions are right for them.

When this will be the case in Europe concerning the ideas of a political union, no one can say for sure. The only thing that is certain is that the influence of committees and resolutions is limited and even the best intentions of the different governments by themselves are in fact necessary but by no means sufficient preconditions. In all probability a number of other components must supervene which have as yet only been realized in outline form:

- The "everyday" Europe must first be developed in some areas and made a matter of course in others, for example as far as border regulations, the exchange of goods, setting up businesses and establishing companies, courses of study and working conditions, taxes and television programs go.
- The "institutional" Europe must lose the stigma of a red tape bureaucracy and of endless bickering and quarreling on a multilingual level.
- The "fundamental" Europe must not fail where, of all places, its presence is most urgently required and where it could offer the most convincing proof for the value of the unification idea – in the control of the emission of harmful substances.
- The "cultural" Europe has to be assured that its historical, ethnic, region-

al, and cultural diversity remains undamaged and untouched, that nothing that is really essential has to be sacrificed on the altar of the delegated authority.

- The "neighbourly" Europe must mean more than simply membership in an organization. Feelings of resentment have to be ignored in coming closer together by means of tourism, youth exchanges and cooperations between regions in different countries.
- The "industry society" Europe has to resolve its most acute structural problems on a domestic basis and not succumb to the only all too enticing temptation of shoving off one's own burdens on the shoulders of the others. Directly related to this is the need to get the differences in social standards under control, at least to the extent that no one can even get the impression that some people are trying to nationalize the advantages of integration while Europeanizing the disadvantages.
- The "information society" Europe has to be sufficiently realized by means of telecommunication networks, joint data banks, the unimpeded flow of data, multilingual satellite programs ("European Culture Channels"), modern transportation links, and a joint development of technology.
- The "geopolitical" Europe has to rediscover its position, in terms of economic and security policy, in the balance of power between East and West and to take on more responsibility in fighting the tragedy of hunger and death in many parts of the Southern Hemisphere.

If one day, let us say, half of these conditions are fulfilled, they will give Europe a new political dimension – perhaps with, perhaps also, though, without official ratifications.

Chapter 20
Reconciliation in the Midst of Contention

Politics Between East and West

Forty years after the end of the war commemoration days, directly or indirectly related to the armistice, followed one upon the other: Four decades ago the division of Europe into spheres of influence was laid down in Yalta, Hitler's Germany surrendered unconditionally, the UNO was founded in San Francisco and the "German question" was discussed in Potsdam. Three decades ago the Federal Republic of Germany joined NATO and established diplomatic relations with the Soviet Union. Fifteen years ago the Moscow and the Warsaw Pact were concluded. Ten years ago the final acts of Helsinki were signed.

Yalta is "an unfinished job," as Zbigniew Brzezinski wrote in an essay in 1984. This is not only true, though, for the consequences of the historical February meeting of the "Big Three" on the Crimean peninsula. A number of resolutions have been reached and commitments entered into between the East and the West since then and they have all contributed to maintaining the peace in Europe. Their merit has been of unique value and cannot be rated highly enough. However, against the background of the continued division of Germany and Europe and the runaway arms race in the world, the maintenance of permanent peace by reconciliation can still be considered "an unfinished job."

Oddly enough, as historical events and points of departure recede into the distance, the more sharply aware people seem to be of this. There is less and less tendency to regard mutual military threat as at least an acceptable, if not an ideal solution, as the concrete memories of worse alternatives, the cold or even the real war, fade. In the eyes of young people what counts above all regarding allied policy is not what it has achieved – avoiding war and safeguarding freedom in Western Europe – but what it should accomplish: disarmament and détente in East-West relations. The generation which has grown up in peace, accustomed to freedom, wants to maintain peace by coming to terms, and not with threats.

It seems that people often forget that the Western defense doctrine has continued unchanged as a double strategy of a credible security policy and a committed policy of détente since the Harmel Report in 1967. It was the Harmel Report which was actually first in introducing the term détente into political usage. The offer to the Soviets to cooperate with them and achieve mutual progress in the area of arms limitation emphasized the solely defensive character of the military aspect of the allied concept. Deterrence was no longer defined, as in the fifties, as a massive nuclear threat of revenge for any kind of attack attempt, but rather as a "flexible response," whose last resort, in the event of all other means failing, could also be the use of atomic weapons.

The alliance did not make things easier for itself with this new strategy. The recognition of a balance of power policy as a stabilizing factor in East-West relations encompassed abstaining from putting the Soviet Union under political pressure by getting a jump on them (which was technologically perfectly possible) in the arms race. The philosophy of détente smoothed the way for a diversive treaty diplomacy which reached its – for the time being – last peak at the summit meeting in Helsinki.

Helsinki also marked, however, the beginnings of a sense of disappointment and disillusionment regarding East-West relations. In the period following the situation of human rights in the Eastern Block countries did not improve anywhere near to the degree many Western conference participants had hoped it would. The Soviet Union began with the installation of new nuclear medium-range missiles which, though not running directly counter to existing treaties, demonstrated disdain for the Western policy of balance because they were not covered by the defense requirements of the Warsaw Pact. The political and military interventions of the USSR in the Middle East, in Africa and especially in Afghanistan in 1979 also contributed towards undermining the spirit and force of the détente policy. Despite everything, however, there still is not a better strategic concept in sight for Europe. Its risks and short-comings have simply become much more apparent. They are due above all to the fact that the term balance of power between the superpowers and their allies incorporates a military and a geopolitical dimension, which the existing treaties do not take fully into account and which can easily be endangered by the many non-European crisis regions. Furthermore, there is the expansive technology push which will be with us for some decades to come and whose effects can scarcely be calculated.

The continued validity of the NATO double strategy does not mean, therefore, that one cannot and even must think about whether the idea of a balance of power through threat of certain destruction can claim validity for all time. Strategic stability, as recent years have clearly demonstrated, is closely related to the question of where the borderline lies between defensive and offensive weapons. Particularly with regard to this question, there are far-reaching differences of opinion between the West and the East. As long as the nuclear offensive potential forms the "hard core" of deterrence, these differences will continue to exist, or even possibly intensify. Should certain threshold countries also someday obtain possession of atomic weapons, and the possibility cannot be ruled out, or other political events weaken the bipolarity of the balance in power, the ability of a security strategy supported by nuclear arms to keep things stable, would begin to vacillate even more.

The pressing question is: What is capable of guaranteeing adequate security in the highly technological, highly sensitive, and highly complex communications world of tomorrow? Technological potentials, moral standards, and the change in economic conditions all have to be taken into consideration.

Thinking out loud for a moment, it seems to me that one cannot completely preclude things developing in the direction of a partial divorce between security and armament standards. On the one hand, the possession of mass annihilation weapons could encounter such strong moral reservations in some Western democracies that the Eastern Block might be tempted into fatal misjudgements, particularly of Western Europe's defence preparedness. From the standpoint of an ideology that tends to judge social orders according to their internal and external strength, our continent then would be a region that was not very security conscious and perhaps willing to gradually switch systems. The nuclear deterrence hypothesis would end up almost having the opposite effect.

On the other hand, and this is also less utopic than it might at first sound, in view of the unavoidable civilian and military dual use of high quality technological products, one can think of certain nations, with technological levels of performance, which would be capable of affording more real protection than conventional weapon arsenals. The more difficult it is to judge the self-defense capacities of a country with modern laser, particle radiation, computer and biotechnology, the less able one is, logically enough, to calculate the risk of an attack.

Still thinking out loud, a completely new security policy constellation could thus one day emerge, with a partition into "two classes of society:" on the one hand the leading technological powers, which, thanks to their know-how, would have enough deterrent "unpredictability potential" at their disposal, and on the other the less technologically developed nations, which would have to try and make up for their shortcomings by arming themselves to the teeth with conventional or atomic weapons because they have no other choice. The bipolarity of the world powers would gradually be transformed into a qualitatively quite different multipolarity: high technology powers with a more defensive as opposed to medium or low technology nations with a more offensive security philosophy.

The former would be closely connected to one other in terms of information and communication technology, but in return would allow for a relatively loose interpretation of political solidarity, with a great deal of autonomy in shaping domestic and foreign policy. The latter, without a doubt at the expense of individual national sovereignty, would have to try and form alliances according to the classical pattern in which those with the greatest force of arms having the greatest political say. The first group would be more apt than the second to evince plurality, individuality, personal freedom, economic and social prosperity, scientific achievements, moral credibility as well as international appeal. A technologically "second-rate" Western Europe would find itself in a quite desperate situation: Against its will, and as far as its strength would allow, it would have to continue to arm itself and would end up falling further and further behind the leading economic and technological nations of the world. Domestically, its various political, social, and ethical standards would be put to a crucial test while its foreign policy would be more and more susceptible to the pressures of trade and power politics from outside forces. In other words, it would become an unstabilized crisis region.

This is simply a scenario, one of many, which are all equally probable or improbable. In asking ourselves what our most efficient, and at the same time acceptable, defense strategies in the future should be, however, it should not be completely ignored. For one thing can certainly now no longer be overlooked: The politicization of high technology, especially under the aspect of security, is in full swing.

In March of 1983 the American president Ronald Reagan made his first public announcement dealing with the space supported Strategic Defense Initiative (SDI) and declared: "What would things be like if the free nations could live safely in the knowledge that their security was not founded on the threat of immediate retaliation in the event of a Soviet attack; that we could intercept and destroy strategic ballistic missiles before they reached our land or those of our allies?" It is now generally agreed that the scientific and technological prerequisites for this "dream of invulnerability" can scarcely be accomplished in the 20th century. In view of the many open questions concerning the programm, one can also not make any final judgements concerning the stabilization effects of its strategy. Nonetheless, the President's question alone demonstrates exactly this new concept of high-tech security. The discussion will not die out, I am convinced, because the

Western democracies will have more and more difficulty in coming to terms with the logical consequences of nuclear power politics, that ultimately rests on being ready to destroy the very thing one wants to defend.

The politicization of high technology will also inevitably result in a new quality in judging traditional forms of trade policy. Manifest economic interests are being increasingly intermixed with legitimate or also just specious security reservations. As Kurt Biedenkopf remarked, in reference to practices involved in the American Cocom, "it is often difficult in a concrete case to distinguish whether one is dealing with a strategic essential or the protection of one's own market."

The "alliance of equals," resting on the two "pillars," the USA and Europe, as John F. Kennedy described it a quarter of a century ago in his famous Philadelphia speech, is headed for a dangerous collision of interests. From the standpoint of security policy, the development of technology should be as parallel and homogeneous as possible on both sides of the Atlantic. From the economic policy standpoint of the stronger partner, at present the USA without a doubt, however, this would mean sustaining losses. This conflict has resulted in a certain tendency towards rigidity and impatience on the one side, which in turn feeds fears and prejudices on the receiving end. The American-European technology transfer is increasingly being used in an inappropriate and unjustifed manner as an index: In the eyes of the one partner it is the litmus test for the "ideological dependability" of the Western Europeans, and the other partner sees it as a proof of whether the USA still wants to deal on a partnership basis in the alliance, or whether it is already acting according to criteria of hegemony.

It is imperative to change course as quickly as possible as the other road is obviously a dead end. Limits to what one can reasonably expect of the other should be paid very close attention to and kept. This limit has undoubtedly been reached, for example, in the case of the Max-Planck Institute for Meteorolgy in Hamburg. A Cyber-205 computer that was ordered in the USA was only delivered after a half year of disputes and massive support by the federal government. Certified end-user documents had to be supplied by the Federal Research Department and precise conditions for the use of the computer were laid down. The rise in the practice of holding American scientific congresses "for US citizens only," understandably also does not exactly arouse friendly feelings in European scientists. Finally, one can expect a certain mininum of understanding for the economic situation in Europe, which is willing to accept security arguments as a motive but not as an excuse for export controls.

On the other hand, one has to acknowledge, from the American standpoint, that certain European modes of behavior impinge on the limit of what one can reasonably expect. When projects like the SDI are subjected to broad moral verdicts in the press and political debate in Western Europe, while at the same time the right to have a share in the relevant research findings is openly claimed, there is hardly anything one can say in response to the accusation that Europe is two-faced.

Countless examples can be found on both sides. One should not dramatize, but we are not dealing with trifles, either. Europe's sovereignty, its ability, at least within the framework of alliance structures that have been established since the war, depends to a large extent on its economic and technological renaissance. Should it fail, far-reaching and incalculable manifestations of disintegration are to be feared – within NATO, the European Community and between individual, particularly differently structured, European nations. There would undoubtedly be a greater temptation, every man for himself, to save what there was to save. In all probability, American politics would tend to center more on the Pacific area while the Soviet Union would undertake massive endeavors to try and win more influence in Western Europe by means of "divide et impera."

No one has more reason to fear this and resist it with all its power than the Federal Republic of Germany. The division of Germany is the crux of European division, and in the contradictory systems of the two German nations the world-wide conflict of ideologies is still concentrated in a particular fashion. For no other country is the dividing line between bridging the gap and falling between two stools, between the chance of reconciliation and the trauma of conflict, between rapprochement and restriction so narrow as in ours. No other country in the West, though, can expect as much advantage from an economically and technologically restrengthened Europe as the free part of Germany: Aside from solving the festering structural crises and cutting down on unemployment, a new sense of political importance could be achieved which could give the European nations a more active role in international committees, in West-East trade and culture exchanges, and in the North-South dialogue.

Periods of confrontation between the great powers which coincide with economic downturns in Western Europe, as we have had to discover and which is only logical in terms of power politics, are periods of the greatest weakness of the Old Continent. In times like these, schemes and embargos, counter schemes and counter embargos flourish, normal economic and cultural contacts between contrary systems are subject to unnatural justifications; negotiations concerning disputes, large or small, stagnate.

Western Europe must and can contribute towards the growth of stability and the elimination of conflict. This includes, above all: being steadfast in the cultivation of economic and cultural contact with the East, without giving cause for any mistrust, and the scientific and technological parity with the powers of the West and the Far East. A strong European pillar in the Atlantic alliance would have a twofold advantage: On the one hand, Europe would be in a better, more convincing position to look after specifically European aspects of security, thus taking some of the load off the USA. On the other hand, Europe's traditional, in terms of the political climate, important role of bridging the gap to the East would be that much more self-evident. Strictly speaking, the continued validity of the alliance strategy – credible defense *and* deténte endeavors – cannot be maintained with a Western Europe that is weak, or worse, domesticated.

The Federal Republic of Germany, without whose forces of economic and technological innovation a new European dynamism seems difficult to imagine, thus has a large responsibility. It extends, if only indirectly, to the nerve centers of the world political crisis management. It stretches, also indirectly, to the coordinates which determine the fate of the divided German peoples, which the two German nations cannot turn about on their own.

For the Federal Republic of Germany and Western Europe meeting the technological challenge is thus much more than simply a matter of economics, a fact which is not always made clear enough in the public debate. The scope of action, in terms of alliance and trade policy, is directly affected by the standard of science and high technology. The less divergency there is between the security and economic interests of Europe and America, the more options – within the framework of the larger political climate – West Germany has in its foreign policy regarding the German question and the Eastern Block. Active cultural and foreign trade contacts between Western and Eastern Europe could in turn also provide new impetus for the flagging détente policy. The West-East dialogue should not be limited to questions of armament alone, as is mainly the case at present. The dominance of military policy in particular only makes matters much worse in achieving progress in the area of arms limitation and disarmament. Technological knowledge, economic power, and political options are always mutually dependent on one another in industrial nations. In no other region, however, are the interests of world policy as closely interrelated as in Western Europe. In its function as bridge-head and intermediate power, Europe has very little choice but to employ its additional power in increased endeavors to achieve a balance between West and East. The United States should also recognize this and judge things accordingly. Technological restrictions against the Western Europeans only end up costing the Americans more in the way of defense expenditures.

The influence that Europe now exercises, and which will perhaps grow in the nineties, can be brought into play, not just in the East, but also in the

South, however, only in the sense of the power to balance differences and alleviate conflicts. It cannot be denied, and the nations of the Third World also acknowledge, that the European nations and the European Community are more serious and unreserved in their commitment to development policy than the great powers and most other non-European industrial countries. While the North-South dialogue generally, and particularly within the forum of the United Nations, is at an almost alltime disastrous low, the Community was able to sign and seal the third Lomé Treaty in December of 1984 within the period stipulated, however drawn out and difficult negotiations might have been. This is essentially a great achievement in the face of the current economic crisis in the world at large, and in Europe in particular, and considering that over 75 nations were involved (the ten member countries of the EC and 65 African, Caribbean, and Pacific developing nations).

And yet Europe could achieve infinitely more if it could manage to get a hold on its own structural problems. Since the conclusion of the Lomé negotiations – the first convention, with at that time 46 African, Caribbean, and Pacific nations, was held in the capital city of Togo in 1975; a decade which unfortunately, but typically hardly registered in the mind of the public – the dispute has centered again and again on problems that all basically boil down to an annex of Europe's unmastered structural transformation. It was and still is a matter of European protection clauses and origin regulations with which the Community can defend itself against the import of manufactured products from the developing countries. It was and still is a matter of the highly subsidized European agricultural surplus exports which drastically reduce the prices for developing countries, whose very existence depends on the profits of their agricultural and raw material exports. Since the Community, together with the USA, practices dumping warfare which cuts down the world market prices even further, it ends up gambling away a good deal of its reputation on the international sugar and grain markets which it has won by negotiations, treaties and exemplary projects.

A Third World policy which is not consistent and willing to make structural adjustments is not credible, and the funds which are made available end up being thrown out the window. The countries of the Third World *must* be able to sell manufactured products to the markets of the industrial nations because it is their only chance of pulling themselves up by their own bootstraps. As high technologies become more sparing of raw material, this type of imports will not be needed by modern industrial nations as much. The result will be a drastic drop in prices, and more and more developing countries, which have lived till now after a fashion from the export of a few, limited raw material monocultures, will end up in completely hopeless situations.

In the last five years the revenues of Third World nations have decreased by an estimated sixty billion dollars due to sinking prices for raw material and agricultural products. With an economic upswing this trend could slow down or even come to a standstill for a time; it cannot be completely turned around, however, in this world of wealthy nations, which have more than they need and which are geared to getting the maximum value out of a minimum use of material. In principle this is not a bad thing: The poorer nations need their natural resources themselves in order to establish a viable economic structure which includes industry, middle-sized businesses, and modern trades. In order to achieve this though, they need information and technology transfer from the Northern Hemisphere, in the form of public financial and training aids, as well as private investments.

A great deal has already been accomplished, but a lot more could be done if there was more political and cultural awareness on both sides. Everything depends on whether the markets in regions where the customers have money to spend are opened up or not. Being realistic, this will only really and truly happen if self-interests are only minimally affected. They are minimal when foreign and domestic products do not overlap much, and they are less apt to overlap the greater the differences are between the underlying basic technologies. As I have already pointed out in the last chapter, nations as well as individuals are under moral obligation to make the most of their cultural, social, and personal talents and potential.

Recently, the thesis has been propagated that things are not in order because the Third World does not need new industries but rather better agricultural structures. Everyone knows that the hunger in the least developed countries is immense, the appalling, soundless dying of millions is beyond the imagination and overtaxes the empathy of even the most committed. What could be more desirable in these tortured and afflicted zones than securing a self-sufficient food supply?

If one really wants to help the farmers of Central Africa to be able to feed themselves, however, one cannot place on them the full burden of getting their countries out of their development ruts. Everyone knows where that leads to: Their yields are meager, and the little that is produced usually ends up being exported because there are no industrial products that can be offered for sale. The little quickly dwindles away, arable land becomes desolate, the desert encroaches without mercy. The people then start dying off or drag themselves to the few large cities, which usually amounts to the same thing, it just takes a little longer. Up until the middle of the last decade, as plants dryed up and the land became parched, the countries in the Sahel zone exported two to five times as much protein as they were able to import in the form of grain. It is estimated that about a fourth of the land in the

Third World is taken up for agricultural imports to the wealthy countries of the world, most of which is used to feed stock. Those who are only concerned with a policy for developing the agriculture, overlook these aspects.

One should also not forget another very important aspect: Before one can start thinking about increasing the yield of basic foodstuffs in many developing countries, the environment first has to gradually, tree for tree, be revived. Since 1945 forest land in the Third World has been halved. Expanses of ground, which all together are several times larger than Europe, are now dead. Till now, this dissolution of nature has been happening on the other side of world, even if we have suddenly been confronted with it in our own forests, though for different reasons. We have better means to combat the dying of the forests and we will use them. In the end though, there is just *one* ecological balance on the earth and it concerns the *one* human race, which can preserve or destroy it.

Europe has the knowledge, the ability, and the power, with the new technologies, to not only help itself, but also others. It can let the Third World take over forms of production and markets which it no longer essentially needs. With training aids and measures for supporting an industrial middle-class, it can contribute towards making the economic structures in developing countries more diverse and stable and thus help to reduce the economic compulsion to overexploit nature. New findings in the area of gene technology can and will have to be placed at the service of projects for developing countries, for example when they succeed in changing plants so that they are capable of thriving in worn out soils. To make Africa's soil fertile again is a biotechnological and environmental vision, which could inspire research just as much as other large-scale industrial or security projects.

The opportunity for progress lies with the Europeans and it is their duty to make the best of it. They are living in one of greatest lands developed and cultivated by man in the world. Their culture is one of heterogeneity within a small space. The heterogeneity can and must be preserved because it will always be an essential element in constituting what cultural Europe is all about. This heterogeneity, however, has also bred an infinite amount of conflicts, which not only the Europeans, but also just about every other nation in the world at some point in their history, have suffered from. When we talk about the values that our continent has bequeathed to the world, we like to envisage ourselves in concert halls, theaters, art galleries and the sanctums of a scientist or a philosopher. This is all true, and there is much of immortal value. But the Europe of colonization, the crusades, the wars, the religious and ideological schisms is also part of the European heritage which we cannot simply shrug off. The Old World has brought forth genius and cruelty, but rarely a sense of community.

We have made a promising start in our conscious effort to discover our sense of community within the framework of heterogeneity. Forty years of peace is, not just in the history of our nations, a long and rare period of time without war and violence. Of the barely more than two dozen nations which, in essence and not just in theory, can be considered free and democratic countries, more than half of them are in our part of the world. Europe could become a political and cultural model of reconciliation if it learns to combine a sense of unity and diversity, to cooperate while preserving its many splendoured identities. Its mission: – after the world has freed itself in the course of the century, first politically, then economically, after that scientifically, and finally technologically, from continental predominance – balance and reconciliation. This calls for political strength, which is not the same thing as power. Part of strength is economic force and this is based, in the era of a new technology, irrevocably on the mastery of this technology. Europe's strength has to go beyond this, however: It is our responsibility to make sure that the voices of the other, prejudiced, tortured, and separated nations are heard and understood in the future information societies. The information age should not be allowed to become a deaf, morally blunted age. What we call progress could otherwise become a retrograde step, segregation and something to regret.

More than the politician, the poet can be ahead of his time, he can be bolder in his assertions. But the politician cannot be denied, with a small ray of hope, quoting the poet, especially if he is one of his own countrymen: "The dissonants of the world are lovers' quarrels" it says in Hölderlin's "Hyperion," "Reconciliation in the midst of contention, and everything divided comes together again."

MIX
Papier aus verantwortungsvollen Quellen
Paper from responsible sources
FSC® C105338

If you have any concerns about our products,
you can contact us on
ProductSafety@springernature.com

In case Publisher is established outside the EU,
the EU authorized representative is:
Springer Nature Customer Service Center GmbH
Europaplatz 3, 69115 Heidelberg, Germany

Printed by Libri Plureos GmbH
in Hamburg, Germany